CAMBRIDGE
UNIVERSITY PRESS

University Printing House, Cambridge, CB2 8BS, United Kingdom

Published in the United States of America by Cambridge University Press, New York

Cambridge University Press is part of the University of Cambridge.
It furthers the University's mission by disseminating knowledge in the pursuit of
education, learning and research at the highest international levels of excellence.

www.cambridge.org
Information on this title: www.cambridge.org/9781108066518

© in this compilation Cambridge University Press 2014

This edition first published 1895
This digitally printed version 2014

ISBN 978-1-108-06651-8 Paperback

opular Scientific Lectures

ERNST MACH

TRANSLATED BY
THOMAS J. MCCORMACK

CAMBRIDGE
UNIVERSITY PRESS

Cambridge University Press has long been a pioneer in the reissuing of out-of-print titles from its own backlist, producing digital reprints of books that are still sought after by scholars and students but could not be reprinted economically using traditional technology. The Cambridge Library Collection extends this activity to a wider range of books which are still of importance to researchers and professionals, either for the source material they contain, or as landmarks in the history of their academic discipline.

Drawing from the world-renowned collections in the Cambridge University Library and other partner libraries, and guided by the advice of experts in each subject area, Cambridge University Press is using state-of-the-art scanning machines in its own Printing House to capture the content of each book selected for inclusion. The files are processed to give a consistently clear, crisp image, and the books finished to the high quality standard for which the Press is recognised around the world. The latest print-on-demand technology ensures that the books will remain available indefinitely, and that orders for single or multiple copies can quickly be supplied.

The Cambridge Library Collection brings back to life books of enduring scholarly value (including out-of-copyright works originally issued by other publishers) across a wide range of disciplines in the humanities and social sciences and in science and technology.

CAMBRIDGE LIBRARY COLLECTION

Books of enduring scholarly value

Physical Sciences

From ancient times, humans have tried to understand the workings of the world around them. The roots of modern physical science go back to the very earliest mechanical devices such as levers and rollers, the mixing of paints and dyes, and the importance of the heavenly bodies in early religious observance and navigation. The physical sciences as we know them today began to emerge as independent academic subjects during the early modern period, in the work of Newton and other 'natural philosophers', and numerous sub-disciplines developed during the centuries that followed. This part of the Cambridge Library Collection is devoted to landmark publications in this area which will be of interest to historians of science concerned with individual scientists, particular discoveries, and advances in scientific method, or with the establishment and development of scientific institutions around the world.

Popular Scientific Lectures

The Austrian scientist Ernst Mach (1838–1916) carried out work of importance in many fields of enquiry, including physics, physiology, psychology and philosophy. Many significant thinkers, such as Ludwig Wittgenstein and Bertrand Russell, benefited from engaging with his ideas. Mach delivered the twelve lectures collected here between 1864 and 1894. This English translation by Thomas J. McCormack (1865–1932) appeared in 1895. Mach tackles a range of topics in an engaging style, demonstrating his abilities as both a researcher and a communicator. In the realm of the physical sciences, he discusses electrostatics, the conservation of energy, and the speed of light. He also addresses physiological matters, seeking to explain aspects of the hearing system and why humans have two eyes. In the final four lectures, he deals with the nature of scientific study. *The Science of Mechanics* (1893), Mach's historical and philosophical account, is also reissued in this series.

POPULAR SCIENTIFIC LECTURES

POPULAR

SCIENTIFIC LECTURES

BY

ERNST MACH

PROFESSOR OF PHYSICS IN THE UNIVERSITY OF PRAGUE

TRANSLATED

BY

THOMAS J. McCORMACK

WITH FORTY-FOUR CUTS AND DIAGRAMS

CHICAGO
THE OPEN COURT PUBLISHING COMPANY
1895

PREFACE.

POPULAR lectures, owing to the knowledge they presuppose, and the time they occupy, can afford only a *modicum* of instruction. They must select for this purpose easy subjects, and restrict themselves to the exposition of the simplest and the most essential points. Nevertheless, by an appropriate choice of the matter, the *charm* and the *poetry* of research can be conveyed by them. It is only necessary to set forth the attractive and the alluring features of a problem, and to show what broad domains of fact can be illuminated by the light radiating from the solution of a single and ofttimes unobtrusive point.

Furthermore, such lectures can exercise a favorable influence by showing the substantial sameness of scientific and every-day thought. The public, in this way, loses its shyness towards scientific questions, and acquires an interest in scientific work which is a great help to the inquirer. The latter, in his turn, is brought to understand that his work is a small part only of the universal process of life, and that the results of his labors must redound to the benefit not only of himself and a few of his associates, but to that of the collective whole.

I sincerely hope that these lectures, in the present excellent translation, will be productive of good in the direction indicated.

E. MACH.

PRAGUE, December, 1894.

TRANSLATOR'S NOTE.

THE present lectures, extending over the period from 1864 to
1894, are here published in *collected* form for the first time.
What few repetitions occur, in the way of examples and quotations,
have been retained, as throwing additional light on the topics they
are designed to illustrate.

As the dates of the first five lectures are not given in the foot-
notes they are here appended. The first lecture, "On the Forms
of Liquids," was delivered in 1868 and published with that "On
Symmetry" in 1872 (Prague). The second and third lectures, on
acoustics, were first published in 1865 (Graz); the fourth and fifth,
on optics, in 1867 (Graz). They belong to the earliest period of
Professor Mach's scientific activity, and with the lectures on electrostatics and education will more than realise the hope expressed in
the author's Preface.

The four remaining lectures are of a more philosophical character and deal principally with the methods and nature of scientific
inquiry. In the ideas summarised in them will be found one of the
most important contributions to the theory of knowledge made in
the last quarter of a century.

All the proofs of this translation have been read by Professor
Mach himself.

T. J. McCormack.

La Salle, Ill., December, 1894.

TABLE OF CONTENTS.

THE FORMS OF LIQUIDS.

WHAT thinkest thou, dear Euthyphron, that the holy is, and the just, and the good? Is the holy holy because the gods love it, or are the gods holy because they love the holy? By such easy questions did the wise Socrates make the market-place of Athens unsafe and relieve presumptuous young statesmen of the burden of imaginary knowledge, by showing them how confused, unclear, and self-contradictory their ideas were.

You know the fate of the importunate questioner. So-called good society avoided him on the promenade. Only the ignorant accompanied him. And finally he drank the cup of hemlock—a lot which we ofttimes wish would fall to modern critics of his stamp.

What we have learned from Socrates, however,— our inheritance from him,— is scientific criticism. Every one who busies himself with science recognises how unsettled and indefinite the notions are which he has brought with him from common life, and how, on a minute examination of things, old differences are effaced and new ones introduced.   The history of science is full of examples of this constant change, development, and clarification of ideas.

But we will not linger by this general consideration of the fluctuating character of ideas, which becomes a source of real uncomfortableness, when we reflect that it applies to almost every notion of life.   Rather shall we observe by the study of a physical example how much a thing changes when it is closely examined, and how it assumes, when thus considered, increasing definiteness of form.

The majority of you think, perhaps, you know quite well the distinction between a liquid and a solid. And precisely persons who have never busied themselves with physics will consider this question one of the easiest that can be put.   But the physicist knows that it is one of the most difficult.   I shall mention here only the experiments of Tresca, which show that solids subjected to high pressures behave exactly as liquids do ; for example, may be made to flow out in the form of jets from orifices in the bottoms of vessels. The supposed difference of kind between liquids and solids is thus shown to be a mere difference of degree.

The common inference that because the earth is oblate in form, it was originally fluid, is an error, in the light of these facts.   True, a rotating sphere, a few inches in diameter will assume an oblate form only if it is very soft, for example, is composed of freshly kneaded clay or some viscous stuff.   But the earth, even if it consisted of the rigidest stone, could not help being crushed by its tremendous weight, and must perforce behave as a fluid. Even our mountains could not extend beyond a certain height without crumbling. The earth *may* once have been fluid, but this by no means follows from its oblateness.

The particles of a liquid are displaced on the application of the slightest pressure; a liquid conforms exactly to the shapes of the vessels in which it is contained; it possesses no form of its own, as you have all learned in the schools. Accommodating itself in the most trifling respects to the conditions of the vessel in which it is placed, and showing, even on its surface, where one would suppose it had the freest play, nothing but a polished, smiling, expressionless countenance, it is the courtier *par excellence* of the natural bodies.

Liquids have no form of their own ! No, not for the superficial observer. But persons who have observed that a raindrop is round and never angular, will not be disposed to accept this dogma so unconditionally.

It is fair to suppose that every man, even the weakest, would possess a character, if it were not too difficult in this world to keep it. So, too, we must suppose that liquids would possess forms of their own, if the pressure of the circumstances permitted it,—if they were not crushed by their own weights.

An astronomer once calculated that human beings could not exist on the sun, apart from its great heat, because they would be crushed to pieces there by their

own weight. The greater mass of this body would
also make the weight of the human body there much
greater.   But on the moon, because here we should
be much lighter, we could jump as high as the church-
steeples without any difficulty, with the same muscular
power which we now possess. Statues and "plaster"
casts of syrup are undoubtedly things of fancy, even
on the moon, but maple-syrup would flow so slowly
there that we could easily build a maple-syrup man on
the moon, for the fun of the thing, just as our children
here build snow-men.

Accordingly, if liquids have no form of their own
with us on earth, they have, perhaps, a form of their
own on the moon, or on some smaller and lighter heavenly body.   The problem, then, simply is to get rid of
the effects of gravity; and, this done, we shall be able
to find out what the peculiar forms of liquids are.

The problem was solved by Plateau of Ghent, whose
method was to immerse the liquid in another of the
same specific gravity.*   He employed for his experiments oil and a mixture of alcohol and water.   By
Archimedes's well-known principle, the oil in this mixture loses its entire weight.   It no longer sinks beneath its weight; its formative forces, be they ever so
weak, are now in full play.

As a fact, we now see, to our surprise, that the oil,
instead of spreading out into a layer, or lying in a

* *Statique expérimentale et théorique des liquides*, 1873. See also *The Science of Mechanics*, p. 384 et seqq.,The Open Court Publishing Co., Chicago, 1893.

formless mass, assumes the shape of a beautiful and perfect sphere, freely suspended in the mixture, as the moon is in space. We can construct in this way a sphere of oil several inches in diameter.

If, now, we affix a thin plate to a wire and insert the plate in the oil sphere, we can, by twisting the wire between our fingers, set the whole ball in rotation. Doing this, the ball assumes an oblate shape, and we can, if we are skilful enough, separate by such rotation a ring from the ball, like that which surrounds Saturn. This ring is finally rent asunder, and, breaking up into a number of smaller balls, exhibits to us a kind of model of the origin of the planetary system according to the hypothesis of Kant and Laplace.

Still more curious are the phenomena exhibited when the formative forces of the liquid are partly disturbed by putting in contact with the liquid's surface some rigid body. If we immerse, for example, the wire framework of a cube in our mass of oil, the oil will everywhere stick to the wire framework. If the quantity of oil is exactly sufficient we shall obtain an oil cube with perfectly smooth walls. If there is too much or too little oil, the walls of the cube will bulge out or cave in. In this manner we

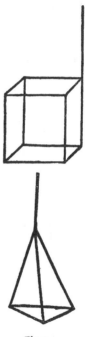

Fig. 1.

can produce all kinds of geometrical figures of oil, for
example, a three-sided pyramid, a cylinder (by bringing the oil between two wire rings), and so on.   Interesting is the change of form that occurs when we
gradually suck out the oil by means of a glass tube
from the cube or pyramid.   The wire holds the oil
fast.   The figure grows smaller and smaller, until it is
at last quite thin.   Ultimately it consists simply of a

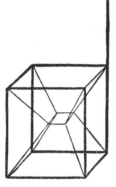 

Fig. 2.

number of thin, smooth plates of oil, which extend
from the edges of the cube to the centre, where they
meet in a small drop. The same is true of the pyramid.

The idea now suggests itself that liquid figures as
thin as this, and possessing, therefore, so slight a
weight, cannot be crushed or deformed by their weight;
just as a small, soft ball of clay is not affected in this
respect by its weight.   This being the case, we no
longer need our mixture of alcohol and water for the
production of figures, but can construct them in the

open air. And Plateau, in fact, found that these thin figures, or at least very similar ones, could be produced in the air, by dipping the wire nets described in a solution of soap and water and quickly drawing them out again. The experiment is not difficult. The figure is formed of itself. The preceding drawing represents to the eye the forms obtained with cubical and pyramidal nets. In the cube, thin, smooth films of soap-suds proceed from the edges to a small, quadratic film in the centre. In the pyramid, a film proceeds from each edge to the centre.

These figures are so beautiful that they hardly admit of appropriate description. Their great regularity and geometrical exactness evokes surprise from all who see them for the first time. Unfortunately, they are of only short duration. They burst, on the drying of the solution in the air, but only after exhibiting to us the most brilliant play of colors, such as is often seen in soap-bubbles. Partly their beauty of form and partly our desire to examine them more minutely induces us to conceive of methods of endowing them with permanent form. This is very simply done.* Instead of dipping the wire nets in solutions of soap, we dip them in pure melted colophonium (resin). When drawn out the figure at once forms and solidifies by contact with the air.

It is to be remarked that also solid fluid-figures can

* Compare Mach, *Ueber die Molecularwirkung der Flüssigkeiten*, Reports of the Vienna Academy, 1862.

be constructed in the open air, if their weight be light enough, or the wire nets of very small dimensions. If we make, for example, of very fine wire a cubical net whose sides measure about one-eighth of an inch in length, we need simply to dip this net in water to obtain a small solid cube of water. With a piece of blotting paper the superfluous water may be easily removed and the sides of the cube made smooth.

Yet another simple method may be devised for observing these figures. A drop of water on a greased glass plate will not run if it is small enough, but will be flattened by its weight, which presses it against its support. The smaller the drop the less the flattening. The smaller the drop the nearer it approaches the form of a sphere. On the other hand, a drop suspended from a stick is elongated by its weight. The undermost parts of a drop of water on a support are pressed against the support, and the upper parts are pressed against the lower parts because the latter cannot yield. But when a drop falls freely downward all its parts move equally fast ; no part is impeded by another ; no part presses against another. A freely falling drop, accordingly, is not affected by its weight ; it acts as if it were weightless ; it assumes a spherical form.

A moment's glance at the soap-film figures produced by our various wire models, reveals to us a great multiplicity of form. But great as this multiplicity is, the common features of the figures also are easily discernible.

"All forms of Nature are allied, though none is the same as the other;
Thus, their common chorus points to a hidden law."

This hidden law Plateau discovered. It may be expressed, somewhat prosily, as follows :

1) If several plane liquid films meet in a figure they are always three in number, and, taken in pairs, form, each with another, nearly equal angles.

2) If several liquid edges meet in a figure they are always four in number, and, taken in pairs, form, each with another, nearly equal angles.

This is a strange law, and its reason is not evident. But we might apply this criticism to almost all laws. It is not always that the motives of a law-maker are discernible in the form of the law he constructs. But our law admits of analysis into very simple elements or reasons. If we closely examine the paragraphs which state it, we shall find that their meaning is simply this, that the surface of the liquid assumes the shape of smallest area that is possible under the circumstances.

If, therefore, some extraordinarily intelligent tailor, possessing a knowledge of all the artifices of the higher mathematics, should set himself the task of so covering the wire frame of a cube with cloth that every piece of cloth should be connected with the wire and joined with the remaining cloth, and should seek to accomplish this feat with the greatest saving of material, he would construct no other figure than that which is here formed on the wire frame in our solution of soap and water. Nature acts in the construction of liquid figures on the principle of a covetous tailor, and gives no thought in her work to the fashions. But, strange to say, in this work, the most beautiful fashions are of themselves produced.

The two paragraphs which state our law apply primarily only to soap-film figures, and are not applicable, of course, to solid oil-figures. But the principle that the superficial area of the liquid shall be the least possible under the circumstances, is applicable to all fluid figures. He who understands not only the letter but also the reason of the law will not be at a loss when confronted with cases to which the letter does not accurately apply. And this is the case with the principle of least superficial area. It is a sure guide for us even in cases in which the above-stated paragraphs are not applicable.

Our first task will now be, to show by a palpable illustration the mode of formation of liquid figures by the principle of least superficial area. The oil on the wire pyramid in our mixture of alcohol and water, being unable to leave the wire edges, clings to them, and the given mass of oil strives so to shape itself that its surface shall have the least possible area. Suppose we attempt to imitate this phenomenon. We take a wire pyramid, draw over it a stout film of rubber, and in place of the wire handle insert a small tube leading

into the interior of the space enclosed by the rubber
(Fig. 3). Through this tube we can blow in or suck
out air. The quantity of air in the enclosure represents the quantity of oil. The stretched rubber film,
which, clinging to the wire edges,
does its utmost to contract, represents the surface of the oil endeavoring to decrease its area. By
blowing in, and drawing out the air,
now, we actually obtain all the oil
pyramidal figures, from those bulged
out to those hollowed in. Finally, when
all the air is pumped or sucked out, the
soap-film figure is exhibited. The rub-

Fig. 3.

ber films strike together, assume the form of planes,
and meet at four sharp edges in the centre of the
pyramid.

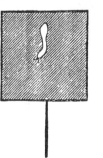 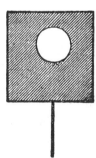

Fig. 4.

The tendency of soap-films to assume smaller forms
may be directly demonstrated by a method of Van der
Mensbrugghe. If we dip a square wire frame to which

a handle is attached into a solution of soap and water, we shall obtain on the frame a beautiful, plane film of soap-suds. (Fig. 4.) On this we lay a thread having its two ends tied together. If, now, we puncture the part enclosed by the thread, we shall obtain a soap-film having a circular hole in it, whose circumference is the thread. The remainder of the film decreasing in area as much as it can, the hole assumes the largest area that it can. But the figure of largest area, with a given periphery, is the circle.

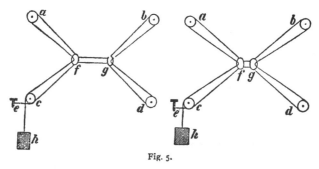

Fig. 5.

Similarly, by the principle of least superficial area, a freely suspended mass of oil assumes the shape of a sphere. The sphere is the form of least surface for a given content. This is evident. The more we put into a travelling-bag, the nearer its shape approaches the spherical form.

The connexion of the two above-mentioned paragraphs with the principle of least superficial area may be shown by a yet simpler example. Picture to yourselves four fixed pulleys, *a, b, c, d,* and two movable

THE FORMS OF LIQUIDS.

rings *f*, *g* (Fig 5); about the pulleys and through the
rings imagine a smooth cord passed, fastened at one
extremity to a nail *e*, and loaded at the other with a
weight *h*. Now this weight always tends to sink, or,
what is the same thing, always tends to make the portion of the string *e h* as long as possible, and consequently the remainder of the string, wound round the
pulleys, as short as possible. The strings must remain
connected with the pulleys, and on account of the rings
also with each other. The conditions of the case, accordingly, are similar to those of the liquid figures discussed. The result also is a similar one. When, as
in the right hand figure of the cut, four pairs of strings
meet, a different configuration must be established.
The consequence of the endeavor of the string to
shorten itself is that the rings separate from each other,
and that now at all points only three pairs of strings
meet, every two at equal angles of one hundred and
twenty degrees. As a fact, by this arrangement the
greatest possible shortening of the string is attained ;
as can be easily proved by geometry.

This will help us to some extent to understand the
creation of beautiful and complicated figures by the
simple tendency of liquids to assume surfaces of least
superficial area. But the question arises, *Why* do
liquids seek surfaces of least superficial area ?

The particles of a liquid cling together. Drops
brought into contact coalesce. We can say, liquid
particles attract each other. If so, they seek to come

as close as they can to each other. The particles at
the surface will endeavor to penetrate as far as they
can into the interior. This process will not stop, cannot stop, until the surface has become as small as under the circumstances it possibly can become, until as
few particles as possible remain at the surface, until
as many particles as possible have penetrated into the
interior, until the forces of attraction have no more
work to perform.*

The root of the principle of least surface is to be
sought, accordingly, in another and much simpler
principle, which may be illustrated by some such analogy as this. We can *conceive* of the natural forces of
attraction and repulsion as purposes or intentions of
nature. As a matter of fact, that interior pressure
which we feel before an act and which we call an intention or purpose, is not, in a final analysis, so essentially different from the pressure of a stone on its support, or the pressure of a magnet on another, that it is
necessarily unallowable to use for both the same term
—at least for well-defined purposes.† It is the purpose of nature, accordingly, to bring the iron nearer
the magnet, the stone nearer the centre of the earth,
and so forth. If such a purpose can be realised, it is
carried out. But where she cannot realise her pur-

---

* In almost all branches of physics that are well worked out such maximal
and minimal problems play an important part.

† Compare Mach, *Vorträge über Psychophysik*, Vienna, 1863, page 41; *Com-
pendium der Physik für Mediciner*, Vienna, 1863, page 234 ; and also *The Science
of Mechanics*, Chicago, 1893, pp. 84 and 464.

poses, nature does nothing. In this respect she acts exactly as a good man of business does.

It is a constant purpose of nature to bring weights lower. We can raise a weight by causing another, larger weight to sink; that is, by satisfying another, more powerful, purpose of nature. If we fancy we are making nature serve our purposes in this, it will be found, upon closer examination, that the contrary is true, and that nature has employed us to attain her purposes.

Equilibrium, rest, exists only, but then always, when nature is brought to a halt in her purposes, when the forces of nature are as fully satisfied as, under the circumstances, they can be. Thus, for example, heavy bodies are in equilibrium, when their so-called centre of gravity lies as low as it possibly can, or when as much weight as the circumstances admit of has sunk as low as it can.

The idea forcibly suggests itself that perhaps this principle also holds good in other realms. Equilibrium exists also in the state when the purposes of the parties are as fully satisfied as for the time being they can be, or, as we may say, jestingly, in the language of physics, when the social potential is a maximum.*

You see, our miserly mercantile principle is replete with consequences.† The result of sober research, it

---

* Like reflexions are found in Quételet, *Du système sociale*.

† For the full development of this idea see the essay "On the Economical Nature of Physical Inquiry," p. 186, and the chapter on "The Economy of Science," in my *Mechanics* (Chicago: The Open Court Publishing Company, 1893), p. 481.

has become as fruitful for physics as the dry questions of Socrates for science generally. If the principle seems to lack in ideality, the more ideal are the fruits which it bears.

But why, tell me, should science be ashamed of such a principle? Is science * itself anything more than—a business? Is not its task to acquire with the least possible work, in the least possible time, with the least possible thought, the greatest possible part of eternal truth?

---

* Science may be regarded as a maximum or minimum problem, exactly as the business of the merchant. In fact, the intellectual activity of natural inquiry is not so greatly different from that exercised in ordinary life as is usually supposed.

THE FIBRES OF CORTI.

WHOEVER has roamed through a beautiful country knows that the tourist's delights increase with his progress. How pretty that wooded dell must look from yonder hill! Whither does that clear brook flow, that hides itself in yonder sedge? If I only knew how the landscape looked behind that mountain! Thus even the child thinks in his first rambles. It is also true of the natural philosopher.

The first questions are forced upon the attention of the inquirer by practical considerations; the subsequent ones are not. An irresistible attraction draws him to these; a nobler interest which far transcends the mere needs of life. Let us look at a special case.

For a long time the structure of the organ of hearing has actively engaged the attention of anatomists. A considerable number of brilliant discoveries has been brought to light by their labors, and a splendid array of facts and truths established. But with these facts a host of new enigmas has been presented.

Whilst in the theory of the organisation and func-

tions of the eye comparative clearness has been attained ; whilst, hand in hand with this, ophthalmology
has reached a degree of perfection which the preceding century could hardly have dreamed of, and by the
help of the ophthalmoscope the observing physician
penetrates into the profoundest recesses of the eye,
the theory of the ear is still much shrouded in mysterious darkness, full of attraction for the investigator.

Look at this model of the ear.   Even at that familiar part by whose extent we measure the quantity of
people's intelligence, even at the external ear, the
problems begin.   You see here a succession of helixes
or spiral windings, at times very pretty, whose significance we cannot accurately state, yet for which there
must certainly þe some reason.

The shell or concha of the ear, *a* in the annexed
diagram, conducts the sound into the curved auditory
passage *b*, which is terminated by a thin membrane,
the so-called tympanic membrane, *e.*   This membrane
is set in motion by the sound, and in its turn sets in
motion a series of little bones of very peculiar formation, *c.*   At the end of all is the labyrinth

d.   The labyrinth consists of a group of
cavities filled with a liquid, in which the

Fig. 6.   innumerable fibres of the nerve of hearing are imbedded.   By the vibration of the chain of
bones *c*, the liquid of the labyrinth is shaken, and the
auditory nerve excited.   Here the process of hearing

begins. So much is certain. But the details of the process are one and all unanswered questions.

To these old puzzles, the Marchese Corti, as late as 1851, added a new enigma. And, strange to say, it is this last enigma, which, perhaps, has first received its correct solution. This will be the subject of our remarks to-day.

Corti found in the cochlea, or snail-shell of the labyrinth, a large number of microscopic fibres placed side by side in geometrically graduated order. According to Kölliker their number is three thousand. They were also the subject of investigation at the hands of Max Schultze and Deiters.

A description of the details of this organ would only weary you, besides not rendering the matter much clearer. I prefer, therefore, to state briefly what in the opinion of prominent investigators like Helmholtz and Fechner is the peculiar function of Corti's fibres. The cochlea, it seems, contains a large number of elastic fibres of graduated lengths (Fig. 7), to which the branches of the auditory nerve are attached. These fibres, called the fibres, pillars, or rods of Corti, being of unequal length, must also be of unequal elasticity, and, consequently, pitched to different notes. The cochlea, therefore, is a species of piano-forte.

Fig. 7.

What, now, may be the office of this structure, which is found in no other organ of sense? May it not be connected with some special property of the ear? It is quite probable; for the ear possesses a very similar power. You know that it is possible to follow the individual voices of a symphony. Indeed, the feat is possible even in a fugue of Bach, where it is certainly no inconsiderable achievement. The ear can pick out the single constituent tonal parts, not only of a harmony, but of the wildest clash of music imaginable. The musical ear analyses every agglomeration of tones.

The eye does not possess this ability. Who, for example, could tell from the mere sight of white, without a previous experimental knowledge of the fact, that white is composed of a mixture of other colors? Could it be, now, that these two facts, the property of the ear just mentioned, and the structure discovered by Corti, are really connected? It is very probable. The enigma is solved if we assume that every note of definite pitch has its special string in this pianoforte of Corti, and, therefore, its special branch of the auditory nerve attached to that string. But before I can make this point perfectly plain to you, I must ask you to follow me a few steps into the dry domain of physics.

Look at this pendulum. Forced from its position of equilibrium by an impulse, it begins to swing with a definite time of oscillation, dependent upon its length. Longer pendulums swing more slowly, shorter ones more quickly. We will suppose our pendulum to execute one to-and-fro movement in a second.

This pendulum, now, can be thrown into violent vibration in two ways; either by a *single* heavy impulse, or by a *number* of properly communicated slight impulses. For example, we impart to the pendulum, while at rest in its position of equilibrium, a very slight impulse. It will execute a very small vibration. As it passes a third time its position of equilibrium, a second having elapsed, we impart to it again a slight shock, in the same direction with the first. Again after the lapse of a second, on its fifth passage through the position of equilibrium, we strike it again in the same manner ; and so continue. You see, by this process the shocks imparted augment continually the motion of the pendulum. After each slight impulse, the pendulum reaches out a little further in its swing, and finally acquires a considerable motion.*

But this is not the case under all circumstances. It is possible only when the impulses imparted synchronise with the swings of the pendulum. If we should communicate the second impulse at the end of half a second and in the same direction with the first impulse, its effects would counteract the motion of the pendulum. It is easily seen that our little impulses help the motion of the pendulum more and more, according as their time accords with the time of the pendulum. If we strike the pendulum in any other time than in that of its vibration, in some instances, it is true, we shall augment its vibration, but in others

* This experiment, with its associated reflexions, is due to Galileo.

again, we shall obstruct it. Our impulses will be less effective the more the motion of our own hand departs from the motion of the pendulum.

What is true of the pendulum holds true of every vibrating body. A tuning-fork when it sounds, also vibrates. It vibrates more rapidly when its sound is higher; more slowly when it is deeper. The standard *A* of our musical scale is produced by about four hundred and fifty vibrations in a second.

I place by the side of each other on this table two tuning-forks, exactly alike, resting on resonant cases. I strike the first one a sharp blow, so that it emits a loud note, and immediately grasp it again with my hand to quench its note. Nevertheless, you still hear the note distinctly sounded, and by feeling it you may convince yourselves that the other fork which was not struck now vibrates.

I now attach a small bit of wax to one of the forks. It is thrown thus out of tune; its note is made a little deeper. I now repeat the same experiment with the two forks, now of unequal pitch, by striking one of them and again grasping it with my hand; but in the present case the note ceases the very instant I touch the fork.

What has happened here in these two experiments? Simply this. The vibrating fork imparts to the air and to the table four hundred and fifty shocks a second, which are carried over to the other fork. If the other fork is pitched to the same note, that is to say, if it vibrates when struck in the same time with the first, then the shocks first emitted, no matter how slight they may be, are sufficient to throw the second fork into rapid sympathetic vibration. But when the time of vibration of the two forks is slightly different, this does not take place. We may strike as many forks as we will, the fork tuned to *A* is perfectly indifferent to their notes; is deaf, in fact, to all except its own ; and if you strike three, or four, or five, or any number whatsoever, of forks all at the same time, so as to make the shocks which come from them ever so great, the *A* fork will not join in with their vibrations unless another fork *A* is found in the collection struck. It picks out, in other words, from all the notes sounded, that which accords with it.

The same is true of all bodies which can yield notes. Tumblers resound when a piano is played, on the striking of certain notes, and so do window panes. Nor is the phenomenon without analogy in other provinces. Take a dog that answers to the name "Nero." He lies under your table. You speak of Domitian, Vespasian, and Marcus Aurelius Antoninus, you call upon all the names of the Roman Emperors that occur to you, but the dog does not stir, although a slight tremor of his ear tells you of a faint response of his consciousness. But the moment you call "Nero" he jumps joyfully towards you. The tuning-fork is like your dog. It answers to the name *A*.

You smile, ladies. You shake your heads. The

simile does not catch your fancy. But I have another, which is very near to you: and for punishment you shall hear it. You, too, are like tuning-forks. Many are the hearts that throb with ardor for you, of which you take no notice, but are cold. Yet what does it profit you ! Soon the heart will come that beats in just the proper rhythm, and then your knell, too, has struck. Then your heart, too, will beat in unison, whether you will or no.

The law of sympathetic vibration, here propounded for sounding bodies, suffers some modification for bodies incompetent to yield notes. Bodies of this kind vibrate to almost every note. A high silk hat, we know, will not sound ; but if you will hold your hat in your hand when attending your next concert you will not only hear the pieces played, but also feel them with your fingers. It is exactly so with men. People who are themselves able to give tone to their surroundings, bother little about the prattle of others. But the person without character tarries everywhere : in the temperance hall, and at the bar of the public-house—everywhere where a committee is formed. The high silk hat is among bells what the weakling is among men of conviction.

A sonorous body, therefore, always sounds when its special note, either alone or in company with others, is struck. We may now go a step further. What will be the behaviour of a group of sonorous bodies which in the pitch of their notes form a scale ? Let us pic-

ture to ourselves, for example (Fig. 8), a series of rods
or strings pitched to the notes *c d e f g*. . . . On a
musical instrument the accord *c e g* is struck. Every
one of the rods of Fig. 8 will see if its special note is
contained in the accord, and if it finds
it, it will respond. The rod *c* will give
at once the note *c*, the rod *e* the note *e*,
the rod *g* the note *g*. All the other
rods will remain at rest, will not sound.
We need not look about us long
for such an instrument. Every piano

c d e f g a b c d e f

Fig. 8.

is an instrument of this kind, with which the experiment mentioned may be executed with splendid success. Two pianos stand here by the side of each other,
both tuned alike. We will employ the first for exciting the notes, while we will allow the second to respond ; after having first pressed upon the loud pedal,
so as to render all the strings capable of motion.

Every harmony struck with vigor on the first piano
is distinctly repeated on the second. To prove that
it is the same strings that are sounded in both pianos,
we repeat the experiment in a slightly changed form.
We let go the loud pedal of the second piano and
pressing on the keys *c e g* of that instrument vigorously
strike the harmony *c e g* on the first piano. The harmony *c e g* is now also sounded on the second piano.
But if we press only on one key *g* of one piano, while
we strike *c e g* on the other, only *g* will be sounded on

the second.   It is thus always the like strings of the
two pianos that excite each other.

The piano can reproduce any sound that is composed of its musical notes.   It will reproduce, for example, very distinctly, a vowel sound that is sung into
it.   And in truth physics has proved that the vowels
may be regarded as composed of simple musical
notes.

You see that by the exciting of definite tones in the
air quite definite motions are set up with mechanical
necessity in the piano.   The idea might be made use
of for the performance of some pretty pieces of wiz-
ardry.   Imagine a box in which is a stretched string
of definite pitch.   This is thrown into motion as often
as its note is sung or whistled.   Now it would not be
a very difficult task for a skilful mechanic to so construct the box that the vibrating cord would close a
galvanic circuit and open the lock.   And it would not
be a much more difficult task to construct a box which
would open at the whistling of a certain melody.   Se-
same ! and the bolts fall.   Truly, we should have here
a veritable puzzle-lock.   Still another fragment rescued from that old kingdom of fables, of which our day
has realised so much, that world of fairy-stories to
which the latest contributions are Casselli's telegraph,
by which one can write at a distance in one's own hand,
and Prof. Elisha Gray's telautograph.   What would
the good old Herodotus have said to these things who
even in Egypt shook his head at much that he saw?

ἐμοὶ μὲν οὐ πιστα, just as simple-heartedly as then, when he heard of the circumnavigation of Africa. A new puzzle-lock! But why invent one? Are not we human beings ourselves puzzle-locks? Think of the stupendous groups of thoughts, feelings, and emotions that can be aroused in us by a word! Are there not moments in all our lives when a mere name drives the blood to our hearts? Who that has attended a large mass-meeting has not experienced what tremendous quantities of energy and motion can be evolved by the innocent words, "Liberty, Equality, Fraternity."

But let us return to the subject proper of our discourse. Let us look again at our piano, or what will do just as well, at some other contrivance of the same character. What does this instrument do? Plainly, it decomposes, it analyses every agglomeration of sounds set up in the air into its individual component parts, each tone being taken up by a different string; it performs a real spectral analysis of sound. A person completely deaf, with the help of a piano, simply by touching the strings or examining their vibrations with a microscope, might investigate the sonorous motion of the air, and pick out the separate tones excited in it.

The ear has the same capacity as this piano. The ear performs for the mind what the piano performs for a person who is deaf. The mind without the ear is deaf. But a deaf person, with the piano, does hear after a fashion, though much less vividly, and more

clumsily, than with the ear.   The ear, thus, also decomposes sound into its component tonal parts.   I shall
now not be deceived, I think, if I assume that you
already have a presentiment of what the function of
Corti's fibres is.   We can make the matter very plain to
ourselves.   We will use the one piano for exciting the
sounds, and we shall imagine the second one in the
ear of the observer in the place of Corti's fibres, which
is a model of such an instrument.   To every string of
the piano in the ear we will suppose a special fibre of
the auditory nerve attached, so that this fibre and this
alone, is irritated when the string is thrown into vibration.   If we strike now an accord on the external
piano, for every tone of that accord a definite string of
the internal piano will sound and as many different
nervous fibres will be irritated as there are notes in
the accord.   The simultaneous sense-impressions due
to different notes can thus be preserved unmingled and
be separated by the attention.   It is the same as with
the five fingers of the hand.   With each finger I can
touch something different.   Now the ear has three thousand such fingers, and each one is designed for the
touching of a different tone.*   Our ear is a puzzle-lock
of the kind mentioned.   It opens at the magic melody
of a sound.   But it is a stupendously ingenious lock.
Not only one tone, but every tone makes it open ; but

* A development of the theory of musical audition differing in many
points from the theory of Helmholtz here expounded, will be found in my
Contributions to the Analysis of the Sensations (English translation by C. M.
Williams), Chicago, The Open Court Publishing Company, 1895.

each one differently. To each tone it replies with a different sensation.

More than once it has happened in the history of science that a phenomenon predicted by theory, has not been brought within the range of actual observation until long afterwards. Leverrier predicted the existence and the place of the planet Neptune, but it was not until sometime later that Galle actually found the planet at the predicted spot. Hamilton unfolded theoretically the phenomenon of the so-called conical refraction of light, but it was reserved for Lloyd some time subsequently to observe the fact. The fortunes of Helmholtz's theory of Corti's fibres have been somewhat similar. This theory, too, received its substantial confirmation from the subsequent observations of V. Hensen. On the free surface of the bodies of Crustacea, connected with the auditory nerves, rows of little hairy filaments of varying lengths and thicknesses are found, which to some extent are the analogues of Corti's fibres. Hensen saw these hairs vibrate when sounds were excited, and when different notes were struck different hairs were set in vibration.

I have compared the work of the physical inquirer to the journey of the tourist. When the tourist ascends a new hill he obtains of the whole district a different view. When the inquirer has found the solution of one enigma, the solution of a host of others falls into his hands.

Surely you have often felt the strange impression experienced when in singing through the scale the octave is reached, and nearly the same sensation is produced as by the fundamental tone. The phenomenon finds its explanation in the view here laid down of the ear. And not only this phenomenon but all the laws of the theory of harmony may be grasped and verified from this point of view with a clearness before undreamt of. Unfortunately, I must content myself to-day with the simple indication of these beautiful prospects. Their consideration would lead us too far aside into the fields of other sciences.

The searcher of nature, too, must restrain himself in his path. He also is drawn along from one beauty to another as the tourist from dale to dale, and as circumstances generally draw men from one condition of life into others. It is not he so much that makes the quests, as that the quests are made of him. Yet let him profit by his time, and let not his glance rove aimlessly hither and thither. For soon the evening sun will shine, and ere he has caught a full glimpse of the wonders close by, a mighty hand will seize him and lead him away into a different world of puzzles.

Respected hearers, science once stood in a different relation to poetry than at present. The old Hindu mathematicians wrote their theorems in verses, and lotus-flowers, roses, and lilies, beautiful sceneries, lakes, and mountains figured in their problems.

"Thou goest forth on this lake in a boat. A lily juts forth, one palm above the water. A breeze bends it downwards, and it vanishes two palms from its previous spot beneath the surface. Quick, mathematician, tell me how deep is the lake ! "

Thus spoke an ancient Hindu scholar. This poetry, and rightly, has disappeared from science, but from its dry leaves another poetry is wafted aloft which cannot be described to him who has never felt it. Whoever will fully enjoy this poetry must put his hand to the plough, must himself investigate. Therefore, enough of this ! I shall reckon myself fortunate if you do not repent of this brief excursion into the flowered dale of physiology, and if you take with yourselves the belief that we can say of science what we say of poetry,

> " Who the song would understand,
> Needs must seek the song's own land;
> Who the minstrel understand
> Needs must seek the minstrel's land."

ON THE CAUSES OF HARMONY.

WE are to speak to-day of a theme which is perhaps of somewhat more general interest—*the causes of the harmony of musical sounds*. The first and simplest experiences relative to harmony are very ancient. Not so the explanation of its laws. These were first supplied by the investigators of a recent epoch. Allow me an historical retrospect.

Pythagoras (586 B. C.) knew that the note yielded by a string of steady tension was converted into its octave when the length of the string was reduced one-half, and into its fifth when reduced two-thirds; and that then the first fundamental tone was consonant with the two others. He knew generally that the same string under fixed tension gives consonant tones when successively divided into lengths that are in the proportions of the simplest natural numbers; that is, in the proportions of 1:2, 2:3, 3:4, 4:5.

Pythagoras failed to reveal the causes of these laws. What have consonant tones to do with the simple natural numbers? That is the question we should ask to-day. But this circumstance must have appeared less strange than inexplicable to Pythagoras. This philosopher sought for the causes of harmony in the occult, miraculous powers of numbers. His procedure was largely the cause of the upgrowth of a numerical mysticism, of which the traces may still be detected in our oneirocritical books and among some scientists, to whom marvels are more attractive than lucidity.

Euclid (300 B. C.) gives a definition of consonance and dissonance that could hardly be improved upon, in point of verbal accuracy. The consonance (συμφωνία) of two tones, he says, is the mixture, the blending (κρᾶσις) of those two tones; dissonance (διαφωνία), on the other hand, is the incapacity of the tones to blend (ἀμιξία), whereby they are made harsh for the ear. The person who knows the correct explanation of the phenomenon hears it, so to speak, reverberated in these words of Euclid. Still, Euclid did not know the true cause of harmony. He had unwittingly come very near to the truth, but without really grasping it.

Leibnitz (1646–1716 A. D.) resumed the question which his predecessors had left unsolved. He, of course, knew that musical notes were produced by vibrations, that twice as many vibrations corresponded to the octave as to the fundamental tone, etc. A passionate lover of mathematics, he sought for the cause of harmony in the secret computation and comparison of the simple numbers of vibrations and in the secret

satisfaction of the soul at this occupation.   But how,
we ask, if one does not know that musical notes are
vibrations ?   The computation and the satisfaction at
the computation must indeed be pretty secret if it is
unknown. What queer ideas philosophers have! Could
anything more wearisome be imagined than computation as a principle of æsthetics ?   Yes, you are not
utterly wrong in your conjecture, yet you may be sure
that Leibnitz's theory is not wholly nonsense, although
it is difficult to make out precisely what he meant by
his secret computation.

The great Euler (1707–1783) sought the cause of
harmony, almost as Leibnitz did, in the pleasure which
the soul derives from the contemplation of order in the
numbers of the vibrations.*

Rameau and D'Alembert (1717–1783) approached
nearer to the truth.   They knew that in every sound
available in music besides the fundamental note also
the twelfth and the next higher third could be heard ;
and further that the resemblance between a fundamental tone and its octave was always strongly marked.
Accordingly, the combination of the octave, fifth, third,
etc., with the fundamental tone appeared to them "natural."   They possessed, we must admit, the correct
point of view ; but with the simple naturalness of a
phenomenon no inquirer can rest content ; for it is pre-

* Sauveur also set out from Leibnitz's idea, but arrived by independent
researches at a different theory, which was very near to that of Helmholtz.
Compare on this point Sauveur, *Mémoires de l'Academie des Sciences*, Paris,
1700-1705, and R. Smith, *Harmonics*, Cambridge, 1749.

cisely this naturalness for which he seeks his explanations.

Rameau's remark dragged along through the whole modern period, but without leading to the full discovery of the truth. Marx places it at the head of his theory of composition, but makes no further application of it. Also Goethe and Zelter in their correspondence were, so to speak, on the brink of the truth. Zelter knew of Rameau's view. Finally, you will be appalled at the difficulty of the problem, when I tell you that till very recent times even professors of physics were dumb when asked what were the causes of harmony.

Not till quite recently did Helmholtz find the solution of the question. But to make this solution clear to you I must first speak of some experimental principles of physics and psychology.

1) In every process of perception, in every observation, the attention plays a highly important part. We need not look about us long for proofs of this. You receive, for example, a letter written in a very poor hand. Do your best, you cannot make it out. You put together now these, now those lines, yet you cannot construct from them a single intelligible character. Not until you direct your attention to groups of lines which really belong together, is the reading of the letter possible. Manuscripts, the letters of which are formed of minute figures and scrolls, can only be read at a considerable distance, where the attention is

no longer diverted from the significant outlines to the
details. A beautiful example of this class is furnished
by the famous iconographs of Giuseppe Arcimboldo in
the basement of the Belvedere gallery at Vienna. These
are symbolic representations of water, fire, etc. : human heads composed of aquatic animals and of combustibles. At a short distance one sees only the details, at a greater distance only the whole figure. Yet
a point can be easily found at which, by a simple voluntary movement of the attention, there is no difficulty
in seeing now the whole figure and now the smaller
forms of which it is composed. A picture is often seen
representing the tomb of Napoleon. The tomb is sur-
rounded by dark trees between which the bright heav-
ens are visible as background. One can look a long time
at this picture without noticing anything except the
trees, but suddenly, on the attention being accidentally directed to the bright background, one sees
the figure of Napoleon between the trees. This case
shows us very distinctly the important part which attention plays. The same sensuous object can, solely
by the interposition of attention, give rise to wholly
different perceptions.

If I strike a harmony, or chord, on this piano, by
a mere effort of attention you can fix every tone of
that harmony. You then hear most distinctly the
fixed tone, and all the rest appear as a mere addition,
altering only the quality, or acoustic color, of the pri-
mary tone. The effect of the same harmony is essen-

tially modified if we direct our attention to different tones.

Strike in succession two harmonies, for example, the two represented in the annexed diagram, and first fix by the attention the upper note *e*, afterwards the base *e—a* ; in the two cases you will hear the same sequence of harmonies differently. In the first case, you have the impression as if the fixed tone remained unchanged and simply altered its *timbre* ; in the second case, the whole acoustic agglomeration seems to fall sensibly in depth.

Fig. 9.

There is an art of composition to guide the attention of the hearer. But there is also an art of hearing, which is not the gift of every person.

The piano-player knows the remarkable effects obtained when one of the keys of a chord that is struck

Fig. 10.

is let loose. Bar 1 played on the piano sounds almost like bar 2. The note which lies next to the key let loose resounds after its release as if it were freshly struck. The attention no longer occupied with the upper note is by that very fact insensibly led to the upper note.

Any tolerably cultivated musical ear can perform the resolution of a harmony into its component parts. By much practice we can go even further.   Then, every musical sound heretofore regarded as simple

Fig. 11.

can be resolved into a subordinate succession of musical tones. For example, if I strike on the piano the note 1, (annexed diagram,) we shall hear, if we make the requisite effort of attention, besides the loud fundamental note the feebler, higher overtones, or harmonics,

2 . . . . 7, that is, the octave, the twelfth, the double octave, and the third, the fifth, and the seventh of the double octave.

The same is true of every musically available sound.   Each yields, with varying degrees of intensity, besides its fundamental note, also the octave, the twelfth, the double octave, etc.   The phenomenon is observable with special facility on the open and closed flue-pipes of organs. According, now, as certain overtones are more or less distinctly emphasised in a sound, the *timbre* of the sound changes—that peculiar quality of the sound by which we distinguish the music of the piano from that of the violin, the clarinet, etc.

On the piano these overtones can be very easily rendered audible. If I strike, for example, sharply note 1 of the foregoing series, whilst I simply press down upon, one after another, the keys 2, 3, ... 7, the notes 2, 3, ... 7 will continue to sound after the

striking of 1, because the strings corresponding to
these notes, now freed from their dampers, are thrown
into sympathetic vibration.

As you know, this sympathetic vibration of the like-
pitched strings with the overtones is really not to be
conceived as sympathy, but rather as lifeless mechanical necessity. We must not think of this sympathetic
vibration as an ingenious journalist pictured it, who
tells a gruesome story of Beethoven's F minor sonata,
Op. 2, that I cannot withhold from you. "At the
last London Industrial Exhibition nineteen virtuosos
played the F minor sonata on the same piano. When
the twentieth stepped up to the instrument to play by
way of variation the same production, to the terror of
all present the piano began to render the sonata of its
own accord. The Archbishop of Canterbury, who
happened to be present, was set to work and forthwith
expelled the F minor devil."

Although, now, the overtones or harmonics which
we have discussed are heard only upon a special effort
of the attention, nevertheless they play a highly important part in the formation of musical *timbre*, as also
in the production of the consonance and dissonance of
sounds. This may strike you as singular. How can
a thing which is heard only under exceptional circumstances be of importance generally for audition?

But consider some familiar incidents of your every-
day life. Think of how many things you see which
you do not notice, which never strike your attention

until they are missing. A friend calls upon you ; you
cannot understand why he looks so changed. Not
until you make a close examination do you discover
that his hair has been cut. It is not difficult to tell
the publisher of a work from its letter-press, and yet
no one can state precisely the points by which this
style of type is so strikingly different from that style.
I have often recognised a book which I was in search
of from a simple piece of unprinted white paper that
peeped out from underneath the heap of books covering it, and yet I had never carefully examined the
paper, nor could I have stated its difference from other
papers.

What we must remember, therefore, is that every
sound that is musically available yields, besides its
fundamental note, its octave, its twelfth, its double
octave, etc., as overtones or harmonics, and that these
are important for the agreeable combination of several
musical sounds.

2) One other fact still remains to be dealt with.
Look at this tuning-fork. It yields, when struck, a perfectly smooth tone. But if you strike in company with
it a second fork which is of slightly different pitch, and
which alone also gives a perfectly smooth tone, you
will hear, if you set both forks on the table, or hold
both before your ear, a uniform tone no longer, but a
number of shocks of tones. The rapidity of the shocks
increases with the difference of the pitch of the forks.
These shocks, which become very disagreeable for the

ear when they amount to thirty-three in a second, are called "beats."

Always, when one of two like musical sounds is thrown out of unison with the other, beats arise. Their number increases with the divergence from unison, and simultaneously they grow more unpleasant. Their roughness reaches its maximum at about thirty-three beats in a second. On a still further departure from unison, and a consequent increase of the number of beats, the unpleasant effect is diminished, so that tones which are widely apart in pitch no longer produce offensive beats.

To give yourselves a clear idea of the production of beats, take two metronomes and set them almost alike. You can, for that matter, set the two exactly alike. You need not fear that they will strike alike. The metronomes usually for sale in the shops are poor enough to yield, when set alike, appreciably unequal strokes. Set, now, these two metronomes, which strike at unequal intervals, in motion ; you will readily see that their strokes alternately coincide and conflict with each other. The alternation is quicker the greater the difference of time of the two metronomes.

If metronomes are not to be had, the experiment may be performed with two watches.

Beats arise in the same way. The rhythmical shocks of two sounding bodies, of unequal pitch, some-times coincide, sometimes interfere, whereby they alternately augment and enfeeble each other's effects. Hence the shock-like, unpleasant swelling of the tone.

Now that we have made ourselves acquainted with overtones and beats, we may proceed to the answer of our main question, Why do certain relations of pitch produce pleasant sounds, consonances, others unpleasant sounds, dissonances? It will be readily seen that all the unpleasant effects of simultaneous sound-combinations are the result of beats produced by those combinations. Beats are the only sin, the sole evil of music. Consonance is the coalescence of sounds without appreciable beats.

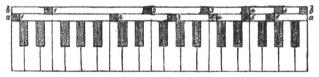

Fig. 12.

To make this perfectly clear to you I have constructed the model which you see in Fig. 12. It represents a claviatur. At its top a movable strip of wood *aa* with the marks 1, 2 . . . . 6 is placed. By setting this strip in any position, for example, in that where the mark 1 is over the note *c* of the claviatur, the marks 2, 3 . . . . 6, as you see, stand over the overtones of *c*. The same happens when the strip is placed in any other position. A second, exactly similiar strip, *bb*, possesses the same properties. Thus, together, the two strips, in any two positions, point out by their

marks all the tones brought into play upon the simultaneous sounding of the notes indicated by the marks 1.

The two strips, placed over the same fundamental
note, show that also all the overtones of those notes
coincide. The first note is simply intensified by the
other. The single overtones of a sound lie too far apart
to permit appreciable beats. The second sound supplies nothing new, consequently, also, no new beats.
Unison is the most perfect consonance.

Moving one of the two strips along the other is
equivalent to a departure from unison. All the overtones of the one sound now fall alongside those of the
other ; beats are at once produced ; the combination
of the tones becomes unpleasant : we obtain a dissonance. If we move the strip further and further along,
we shall find that as a general rule the overtones always fall alongside each other, that is, always produce
beats and dissonances. Only in a few quite definite
positions do the overtones partially coincide. Such
positions, therefore, signify higher degrees of euphony
—they point out *the consonant intervals.*

These consonant intervals can be readily found experimentally by cutting Fig. 12 out of paper and moving
bb lengthwise along *aa*. The most perfect consonances
are the octave and the twelfth, since in these two cases
the overtones of the one sound coincide absolutely
with those of the other. In the octave, for example,
1 *b* falls on 2 *a*, 2 *b* on 4 *a*, 3 *b* on 6 *a*. Consonances,
therefore, are simultaneous sound-combinations not

accompanied by disagreeable beats. This, by the way, is, expressed in English, what Euclid said in Greek.

Only such sounds are consonant as possess in common some portion of their partial tones. Plainly we must recognise between such sounds, also when struck one after another, a certain affinity. For the second sound, by virtue of the common overtones, will produce partly the same sensation as the first. The octave is the most striking exemplification of this. When we reach the octave in the ascent of the scale we actually fancy we hear the fundamental tone repeated. The foundations of harmony, therefore, are the foundations of melody.

Consonance is the coalescence of sounds without appreciable beats ! This principle is competent to introduce wonderful order and logic into the doctrines of the fundamental bass. The compendiums of the theory of harmony which (Heaven be witness !) have stood hitherto little behind the cook-books in subtlety of logic, are rendered extraordinarily clear and simple. And what is more, all that the great masters, such as Palestrina, Mozart, Beethoven, unconsciously got right, and of which heretofore no text-book could render just account, receives from the preceding principle its perfect verification.

But the beauty of the theory is, that it bears upon its face the stamp of truth. It is no phantom of the brain. Every musician can hear for himself the beats which the overtones of his musical sounds produce.

Every musician can satisfy himself that for any given case the number and the harshness of the beats can be calculated beforehand, and that they occur in exactly the measure that theory determines.

This is the answer which Helmholtz gave to the question of Pythagoras, so far as it can be explained with the means now at my command. A long period of time lies between the raising and the solving of this question. More than once were eminent inquirers nearer to the answer than they dreamed of.

The inquirer seeks the truth. I do not know if the truth seeks the inquirer. But were that so, then the history of science would vividly remind us of that classical rendezvous, so often immortalised by painters and poets. A high garden wall. At the right a youth, at the left a maiden. The youth sighs, the maiden sighs! Both wait. Neither dreams how near the other is.

I like this simile. Truth suffers herself to be courted, but she has evidently no desire to be won. She flirts at times disgracefully. Above all, she is determined to be merited, and has naught but contempt for the man who will win her too quickly. And if, forsooth, one breaks his head in his efforts of conquest, what matter is it, another will come, and truth is always young. At times, indeed, it really seems as if she were well disposed towards her admirer, but that admitted—never! Only when Truth is in exceptionally good spirits does she bestow upon her wooer a glance of encouragement. For, thinks Truth, if I do not do something, in the end the fellow will not seek me at all.

This one fragment of truth, then, we have, and it shall never escape us. But when I reflect what it has cost in labor and in the lives of thinking men, how it painfully groped its way through centuries, a half-matured thought, before it became complete; when I reflect that it is the toil of more than two thousand years that speaks out of this unobtrusive model of mine, then, without dissimulation, I almost repent me of the jest I have made.

And think of how much we still lack! When, several thousand years hence, boots, top-hats, hoops, pianos, and bass-viols are dug out of the earth, out of the newest alluvium as fossils of the nineteenth century; when the scientists of that time shall pursue their studies both upon these wonderful structures and upon our modern Broadways, as we to-day make studies of the implements of the stone age and of the prehistoric lake-dwellings—then, too, perhaps, people will be unable to comprehend how we could come so near to many great truths without grasping them. And thus it is for all time the unsolved dissonance, for all time the troublesome seventh, that everywhere resounds in our ears; we feel, perhaps, that it will find its solution, but we shall never live to see the day of the pure triple accord, nor shall our remotest descendants.

Ladies, if it is the sweet purpose of your life to sow confusion, it is the purpose of mine to be clear;

and so I must confess to you a slight transgression that I have been guilty of. On one point I have told you an untruth. But you will pardon me this false-hood, if in full repentance I make it good. The model represented in Fig. 12 does not tell the whole truth, for it is based upon the so-called "even temperament" system of tuning. The overtones, however, of musical sounds are not tempered, but purely tuned. By means of this slight inexactness the model is made consider-ably simpler. In this form it is fully adequate for ordinary purposes, and no one who makes use of it in his studies need be in fear of appreciable error.

If you should demand of me, however, the full truth, I could give you that only by the help of a math-ematical formula. I should have to take the chalk into my hands and—think of it !—reckon in your presence. This you might take amiss. Nor shall it happen. I have resolved to do no more reckoning for to-day. I shall reckon now only upon your forbearance, and this you will surely not gainsay me when you reflect that I have made only a limited use of my privilege to weary you. I could have taken up much more of your time, and may, therefore, justly close with Les-sing's epigram :

" If thou hast found in all these pages naught that's worth the thanks,
At least have gratitude for what I've spared thee."

THE VELOCITY OF LIGHT.

WHEN a criminal judge has a right crafty knave before him, one well versed in the arts of prevarication, his main object is to wring a confession from the culprit by a few skilful questions. In almost a similar position the natural philosopher seems to be placed with respect to nature. True, his functions here are more those of the spy than the judge; but his object remains pretty much the same. Her hidden motives and laws of action is what nature must be made to confess. Whether a confession will be extracted depends upon the shrewdness of the inquirer. Not without reason, therefore, did Lord Bacon call the experimental method a questioning of nature. The art consists in so putting our questions that they may not remain unanswered without a breach of etiquette.

Look, too, at the countless tools, engines, and instruments of torture with which man conducts his inquisitions of nature, and which mock the poet's words :

" Mysterious even in open day,
Nature retains her veil, despite our clamors;
That which she doth not willingly display
Cannot be wrenched from her with levers, screws, and hammers."

Look at these instruments and you will see that the comparison with torture also is admissible.* This view of nature, as of something designedly concealed from man, that can be unveiled only by force or dishonesty, chimed in better with the conceptions of the ancients than with modern notions. A Grecian philosopher once said, in offering his opinion of the natural science of his time, that it could only be displeasing to the gods to see men endeavoring to spy out what the gods were not minded to reveal to them.† Of course all the contemporaries of the speaker were not of his opinion.

Traces of this view may still be found to-day, but upon the whole we are now not so narrow-minded. We believe no longer that nature designedly hides herself. We know now from the history of science that our questions are sometimes meaningless, and that, therefore, no answer can be forthcoming. Soon we shall see how man, with all his thoughts and quests, is only a fragment of nature's life.

---

* According to Mr. Jules Andrieu, the idea that nature must be tortured to reveal her secrets is preserved in the name *crucible*—from the Latin *crux*, a cross. But, more probably, *crucible* is derived from some Old French or Teutonic form, as *cruche*, *kroes*, *krus*, etc., a pot or jug (cf. Modern English *crock*, *cruse*, and German *Krug*).—*Trans.*

† Xenophon, Memorabilia iv, 7, puts into the mouth of Socrates these words : οὔτε γὰρ εὑρετὰ ἀνθρώποις αὐτὰ ἐνόμιζεν εἶναι, οὔτε χαοίζεσθαι θεοῖς ἂν ἡγεῖτο τὸν ζητοῦντα ἃ ἐκεῖνοι σαφηνίσαι οὐκ ἐβουλήθησαν.

Picture, then, as your fancy dictates, the tools of the physicist as instruments of torture or as engines of endearment, at all events a chapter from the history of those implements will be of interest to you, and it will not be unpleasant to learn what were the peculiar difficulties that led to the invention of such strange apparatus.

Galileo (born at Pisa in 1564, died at Arcetri in 1642) was the first who asked what was the velocity of light, that is, what time it would take for a light struck at one place to become visible at another, a certain distance away.*

The method which Galileo devised was as simple as it was natural. Two practised observers, with muffled lanterns, were to take up positions in a dark night at a considerable dis-

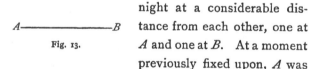

Fig. 13.

tance from each other, one at A and one at B. At a moment previously fixed upon, A was instructed to unmask his lantern; while as soon as B saw the light of A's lantern he was to unmask his. Now it is clear that the time which A counted from the uncovering of his lantern until he caught sight of the light of B's would be the time which it would take light to travel from A to B and from B back to A.

The experiment was not executed, nor could it, in the nature of the case, have been a success. As we

---

* Galilei, *Discorsi e dimostrazione matematiche.* Leyden, 1638. *Dialogo Primo.*

now know, light travels too rapidly to be thus noted. The time elapsing between the arrival of the light at *B* and its perception by the observer, with that between the decision to uncover and the uncovering of the lantern, is, as we now know, incomparably greater than the time which it takes light to travel the greatest earthly distances. The great velocity of light will be made apparent, if we reflect that a flash of lightning in the night illuminates instantaneously a very extensive region, whilst the single reflected claps of thunder arrive at the observer's ear very gradually and in appreciable succession.

During his life, then, the efforts of Galileo to determine the velocity of light remained uncrowned with success. But the subsequent history of the measurement of the velocity of light is intimately associated with his name, for with the telescope which he constructed he discovered the four satellites of Jupiter, and these furnished the next occasion for the determination of the velocity of light.

The terrestrial spaces were too small for Galileo's experiment. The measurement was first executed when the spaces of the planetary system were employed. Olaf Römer, (born at Aarhuus in 1644, died at Copenhagen in 1710) accomplished the feat (1675–1676), while watching with Cassini at the observatory of Paris the revolutions of Jupiter's moons.

Let *AB* (Fig. 14) be Jupiter's orbit. Let *S* stand for the sun, *E* for the earth, *J* for Jupiter, and *T* for

Jupiter's first satellite. When the earth is at E_1 we
see the satellite enter regularly into Jupiter's shadow,
and by watching the time between two successive
eclipses, can calculate its time of revolution. The
time which Römer noted was forty-two hours, twenty-
eight minutes, and thirty-five seconds. Now, as the
earth passes along in its orbit towards E_2, the revolutions of the satellite grow apparently longer and longer :

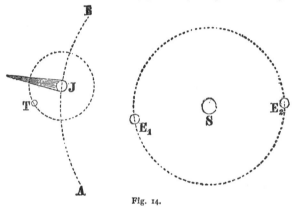

Fig. 14.

the eclipses take place later and later. The greatest
retardation of the eclipse, which occurs when the earth
is at E_2, amounts to sixteen minutes and twenty-six
seconds. As the earth passes back again to E_1, the
revolutions grow apparently shorter, and they occur
in exactly the time that they first did when the earth
arrives at E_1. It is to be remarked that Jupiter changes
only very slightly its position during one revolution of
the earth. Römer guessed at once that these periodical changes of the time of revolution of Jupiter's satel-

lite were not actual, but apparent changes, which were in some way connected with the velocity of light.

Let us make this matter clear to ourselves by a simile. We receive regularly by the post, news of the political status at our capital. However far away we may be from the capital, we hear the news of every event, later it is true, but of all equally late. The events reach us in the same succession of time as that in which they took place. But if we are travelling away from the capital, every successive post will have a greater distance to pass over, and the events will reach us more slowly than they took place. The reverse will be the case if we are approaching the capital.

At rest, we hear a piece of music played in the same *tempo* at all distances. But the *tempo* will be seemingly accelerated if we are carried rapidly towards the band, or retarded if we are carried rapidly away from it.*

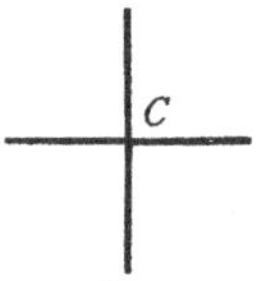

Fig. 15.

Picture to yourself a cross, say the sails of a wind-mill (Fig. 15), in uniform rotation about its centre. Clearly, the rotation of the cross will appear to you more slowly executed if you are carried very rapidly away from it. For the post which in this case conveys to you the light and brings to you the news of the successive positions of the cross will have to travel in each successive instant over a longer path.

*In the same way, the pitch of a locomotive-whistle is higher as the locomotive rapidly approaches an observer, and lower when rapidly leaving him than if the locomotive were at rest.—*Trans.*

Now this must also be the case with the rotation (the revolution) of the satellite of Jupiter. The greatest retardation of the eclipse (16½ minutes), due to the passage of the earth from E_1 to E_2, or to its removal from Jupiter by a distance equal to the diameter of the orbit of the earth, plainly corresponds to the time which it takes light to traverse a distance equal to the diameter of the earth's orbit. The velocity of light, that is, the distance described by light in a second, as determined by this calculation, is 311,000 kilometres,[*] or 193,000 miles. A subsequent correction of the diameter of the earth's orbit, gives, by the same method, the velocity of light as approximately 186,000 miles a second.

The method is exactly that of Galileo; only better conditions are selected. Instead of a short terrestrial distance we have the diameter of the earth's orbit, three hundred and seven million kilometres; in place of the uncovered and covered lanterns we have the satellite of Jupiter, which alternately appears and disappears. Galileo, therefore, although he could not carry out himself the proposed measurement, found the lantern by which it was ultimately executed.

Physicists did not long remain satisfied with this beautiful discovery. They sought after easier methods of measuring the velocity of light, such as might be performed on the earth. This was possible after the difficulties of the problem were clearly exposed. A

---

[*] A kilometre is 0.621 or nearly five-eighths of a statute mile.

measurement of the kind referred to was executed in
1849 by Fizeau (born at Paris in 1819).

I shall endeavor to make the principle of Fizeau's
apparatus clear to you. Let *s* (Fig. 16) be a disk free
to rotate about its centre, and perforated at its rim
with a series of holes. Let *l* be a luminous point
casting its light on an unsilvered glass, *a*, inclined at
an angle of forty-five degrees to the axis of the disk.
The ray of light, reflected at this point, passes through
one of the holes of the disk and falls at right angles

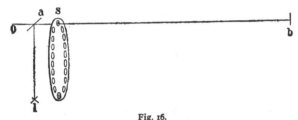

Fig. 16.

upon a mirror *b*, erected at a point about five miles
distant. From the mirror *b* the light is again reflected,
passes once more through the hole in *s*, and, penetrat-
ing the glass plate, finally strikes the eye, *o*, of the observer. The eye, *o*, thus, sees the image of the luminous point *l* through the glass plate and the hole of
the disk in the mirror *b*.

If, now, the disk be set in rotation, the unpierced
spaces between the apertures will alternately take the
place of the apertures, and the eye *o* will now see the
image of the luminous point in *b* only at interrupted
intervals. On increasing the rapidity of the rotation,

however, the interruptions for the eye become again
unnoticeable, and the eye sees the mirror *b* uniformly
illuminated.

But all this holds true only for relatively small velocities of the disk, when the light sent through an
aperture in *s* to *b* on its return strikes the aperture at
almost the same place and passes through it a second
time. Conceive, now, the speed of the disk to be so increased that the light on its return finds before it an
unpierced space instead of an aperture, it will then no
longer be able to reach the eye. We then see the
mirror *b* only when no light is emitted from it, but
only when light is sent to it ; it is covered when light
comes from it. In this case, accordingly, the mirror
will always appear dark.

If the velocity of rotation at this point were still
further increased, the light sent through one aperture
could not, of course, on its return pass through the
same aperture but might strike the next and reach
the eye by that. Hence, by constantly increasing the
velocity of the rotation, the mirror *b* may be made to
appear alternately bright and dark. Plainly, now, if
we know the number of apertures of the disk, the number of rotations per second, and the distance *s b*, we
can calculate the velocity of light. The result agrees
with that obtained by Römer.

The experiment is not quite as simple as my exposition might lead you to believe. Care must be
taken that the light shall travel back and forth over

the miles of distance *s b* and *b s* undispersed. This difficulty is obviated by means of telescopes.

If we examine Fizeau's apparatus closely, we shall recognise in it an old acquaintance : the arrangement of Galileo's experiment. The luminous point *l* is the lantern *A*, while the rotation of the perforated disk performs mechanically the uncovering and covering of the lantern. Instead of the unskilful observer *B* we have the mirror *b*, which is unfailingly illuminated the instant the light arrives from *s*. The disk *s*, by alternately transmitting and intercepting the reflected light, assists the observer *o*. Galileo's experiment is here executed, so to speak, countless times in a second, yet the total result admits of actual observation. If I might be pardoned the use of a phrase of Darwin's in this field, I should say that Fizeau's apparatus was the descendant of Galileo's lantern.

A still more refined and delicate method for the measurement of the velocity of light was employed by Foucault, but a description of it here would lead us too far from our subject.

The measurement of the velocity of sound is easily executed by the method of Galileo. It was unnecessary, therefore, for physicists to rack their brains further about the matter ; but the idea which with light grew out of necessity was applied also in this field. Koenig of Paris constructs an apparatus for the measurement of the velocity of sound which is closely allied to the method of Fizeau.

The apparatus is very simple. It consists of two electrical clock-works which strike simultaneously, with perfect precision, tenths of seconds. If we place the two clock-works directly side by side, we hear their strokes simultaneously, wherever we stand. But if we take our stand by the side of one of the works and place the other at some distance from us, in general a coincidence of the strokes will now not be heard. The companion strokes of the remote clock-work arrive, as sound, later. The first stroke of the remote work is heard, for example, immediately after the first of the adjacent work, and so on. But by increasing the distance we may produce again a coincidence of the strokes. For example, the first stroke of the remote work coincides with the second of the near work, the second of the remote work with the third of the near work, and so on. If, now, the works strike tenths of seconds and the distance between them is increased until the first coincidence is noted, plainly that distance is travelled over by the sound in a tenth of a second.

We meet frequently the phenomenon here presented, that a thought which centuries of slow and painful endeavor are necessary to produce, when once developed, fairly thrives. It spreads and runs everywhere, even entering minds in which it could never have arisen. It simply cannot be eradicated.

The determination of the velocity of light is not the only case in which the direct perception of the senses is too slow and clumsy for use. The usual method of studying events too fleet for direct observation consists in putting into reciprocal action with them other events already known, the velocities of all of which are capable of comparison. The result is usually unmistakable, and susceptible of direct inference respecting the character of the event which is unknown. The velocity of electricity cannot be determined by direct observation. But it was ascertained by Wheatstone, simply by the expedient of watching an electric spark in a mirror rotating with tremendous known velocity.

Fig. 17.

If we wave a staff irregularly hither and thither, simple observation cannot determine how quickly it moves at each point of its course. But let us look at the staff through holes in the rim of a rapidly rotating disk (Fig. 17). We shall then see the moving staff only in certain positions, namely, when a hole passes in front of the eye. The single pictures of the staff remain for a time impressed upon the eye ; we think we see several staffs, having some such disposition as that represented in Fig. 18. If, now, the holes of the disk are equally far apart, and the disk is rotated with uniform velocity, we see clearly that the staff has moved slowly from *a* to *b*, more quickly from *b* to *c*, still more quickly from *c* to *d*, and with its greatest velocity from *d* to *e*.

Fig. 18.

A jet of water flowing from an orifice in the bottom of a vessel has the appearance of perfect quiet and uniformity, but if we illuminate it for a second, in a dark room, by means of an electric flash we shall see that the jet is composed of separate drops.   By their quick descent the images of the drops are obliterated and the jet appears uniform. Let us look at the jet through the rotating disk.   The disk is supposed to be rotated so rapidly that while the second aperture passes into the place of the first, drop 1 falls into the place of 2, 2 into the place of 3, and so on. We see drops then always in the same places.   The jet appears to be at rest.   If we turn the disk a trifle more slowly, then while the second aperture passes into the place of the first, drop 1 will have fallen somewhat lower than 2, 2 somewhat lower than 3, etc. Through every successive aperture we shall see drops in successively lower positions.   The jet will appear to be flowing slowly downwards.

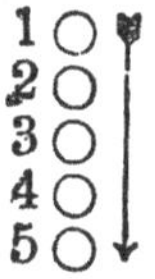

Now let us turn the disk more rapidly.   Then while the second aperture is passing into the place of the first, drop 1 will not quite have reached the place of 2, but will be found slightly above 2, 2 slightly above 3, etc.   Through the successive apertures we shall see the drops at successively higher places.   It will now look as if the jet were flowing upwards, as if the drops were rising from the lower vessel into the higher.

You see, physics grows gradually more and more

Fig. 19.

1 2 3 4 5

terrible. The physicist will soon have it in his power to play the part of the famous lobster chained to the bottom of the Lake of Mohrin, whose direful mission, if ever liberated, the poet Kopisch humorously describes as that of a reversal of all the events of the world; the rafters of houses become trees again, cows calves, honey flowers, chickens eggs, and the poet's own poem flows back into his inkstand.

* * *

You will now allow me the privilege of a few general remarks. You have seen that the same principle often lies at the basis of large classes of apparatus designed for different purposes. Frequently it is some very unobtrusive idea which is productive of so much fruit and of such extensive transformations in physical technics. It is not different here than in practical life.

The wheel of a waggon appears to us a very simple and insignificant creation. But its inventor was certainly a man of genius. The round trunk of a tree perhaps first accidentally led to the observation of the ease with which a load can be moved on a roller. Now, the step from a simple supporting roller to a fixed roller, or wheel, appears a very easy one. At least it appears very easy to us who are accustomed from childhood up to the action of the wheel. But if we put ourselves vividly into the position of a man who never saw a wheel, but had to invent one, we shall begin to have some idea of its difficulties. Indeed, it

is even doubtful whether a single man could have accomplished this feat, whether perhaps centuries were
not necessary to form the first wheel from the primitive roller.*

History does not name the progressive minds who
constructed the first wheel; their time lies far back of
the historic period. No scientific academy crowned
their efforts, no society of engineers elected them
honorary members. They still live only in the stupendous results which they called forth. Take from
us the wheel, and little will remain of the arts and industries of modern life. All disappears. From the
spinning-wheel to the spinning-mill, from the turning-
lathe to the rolling-mill, from the wheelbarrow to the
railway train, all vanishes.

In science the wheel is equally important. Whirl-
ing machines, as the simplest means of obtaining quick
motions with inconsiderable changes of place, play a
part in all branches of physics. You know Wheatstone's rotating mirror, Fizeau's wheel, Plateau's perforated rotating disks, etc. Almost the same principle
lies at the basis of all these apparatus. They differ
from one another no more than the pen-knife differs,
in the purposes it serves, from the knife of the anatomist or the knife of the vine-dresser. Almost the same
might be said of the screw.

---

*Observe, also, the respect in which the wheel is held in India, Japan
and other Buddhistic countries, as the emblem of power, order, and law, and
of the superiority of mind over matter. The consciousness of the importance of
this invention seems to have lingered long in the minds of these nations.—*Tr.*

It will now perhaps be clear to you that new thoughts do not spring up suddenly. Thoughts need their time to ripen, grow, and develop in, like every natural product; for man, with his thoughts, is also a part of nature.

Slowly, gradually, and laboriously one thought is transformed into a different thought, as in all likelihood one animal species is gradually transformed into new species. Many ideas arise simultaneously. They fight the battle for existence not differently than do the Ichthyosaurus, the Brahman, and the horse.

A few remain to spread rapidly over all fields of knowledge, to be redeveloped, to be again split up, to begin again the struggle from the start. As many animal species long since conquered, the relics of ages past, still live in remote regions where their enemies cannot reach them, so also we find conquered ideas still living on in the minds of many men. Whoever will look carefully into his own soul will acknowledge that thoughts battle as obstinately for existence as animals. Who will gainsay that many vanquished modes of thought still haunt obscure crannies of his brain, too faint-hearted to step out into the clear light of reason? What inquirer does not know that the hardest battle, in the transformation of his ideas, is fought with himself.

Similar phenomena meet the natural inquirer in all paths and in the most trifling matters. The true inquirer seeks the truth everywhere, in his country-

walks and on the streets of the great city. If he is
not too learned, he will observe that certain things,
like ladies' hats, are constantly subject to change. I
have not pursued special studies on this subject, but
as long as I can remember, one form has always
gradually changed into another. First, they wore hats
with long projecting rims, within which, scarcely accessible with a telescope, lay concealed the face of the
beautiful wearer. The rim grew smaller and smaller;
the bonnet shrank to the irony of a hat. Now a tremendous superstructure is beginning to grow up in its
place, and the gods only know what its limits will be.
It is not different with ladies' hats than with butter-
flies, whose multiplicity of form often simply comes
from a slight excrescence on the wing of one species
developing in a cognate species to a tremendous fold.
Nature, too, has its fashions, but they last thousands
of years. I could elucidate this idea by many additional examples; for instance, by the history of the
evolution of the coat, if I were not fearful that my
gossip might prove irksome to you.

*   *   *

We have now wandered through an odd corner of
the history of science. What have we learned? The
solution of a small, I might almost say insignificant,
problem—the measurement of the velocity of light.
And more than two centuries have worked at its solution! Three of the most eminent natural philosophers,
Galileo, an Italian, Römer, a Dane, and Fizeau, a

Frenchman, have fairly shared its labors. And so it is with countless other questions. When we contemplate thus the many blossoms of thought that must wither and fall before one shall bloom, then shall we first truly appreciate Christ's weighty but little consolatory words : " Many be called but few are chosen."

Such is the testimony of every page of history. But is history right? Are really only those chosen whom she names ? Have those lived and battled in vain, who have won no prize?

I doubt it. And so will every one who has felt the pangs of sleepless nights spent in thought, at first fruitless, but in the end successful. No thought in such struggles was thought in vain ; each one, even the most insignificant, nay, even the erroneous thought, that which apparently was the least productive, served to prepare the way for those that afterwards bore fruit. And as in the thought of the individual naught is in vain, so, also, it is in that of humanity.

Galileo wished to measure the velocity of light. He had to close his eyes before his wish was realised. But he at least found the lantern by which his successor could accomplish the task.

And so I may maintain that we all, so far as inclination goes, are working at the civilisation of the future. If only we all strive for the right, then are we *all* called and *all* chosen !

WHY HAS MAN TWO EYES?

WHY has man two eyes? That the pretty symmetry of his face may not be disturbed, the artist answers. That his second eye may furnish a substitute for his first if that be lost, says the far-sighted economist. That we may weep with two eyes at the sins of the world, replies the religious enthusiast.

Odd opinions! Yet if you should approach a modern scientist with this question you might consider yourself fortunate if you escaped with less than a rebuff. "Pardon me, madam, or my dear sir," he would say, with stern expression, "man fulfils no purpose in the possession of his eyes; nature is not a person, and consequently not so vulgar as to pursue purposes of any kind."

Still an unsatisfactory answer! I once knew a professor who would shut with horror the mouths of his pupils if they put to him such an unscientific question.

But ask a more tolerant person, ask me. I, I candidly confess, do not know exactly why man has two eyes, but the reason partly is, I think, that I may see you here before me to-night and talk with you upon this delightful subject.

Again you smile incredulously. Now this is one of those questions that a hundred wise men together could not answer. You have heard, so far, only five of these wise men. You will certainly want to be spared the opinions of the other ninety-five. To the first you will reply that we should look just as pretty if we were born with only one eye, like the Cyclops ; to the second we should be much better off, according to his principle, if we had four or eight eyes, and that in this respect we are vastly inferior to spiders ; to the third, that you are not just in the mood to weep ; to the fourth, that the unqualified interdiction of the question excites rather than satisfies your curiosity ; while of me you will dispose by saying that my pleasure is not as intense as I think, and certainly not great enough to justify the existence of a double eye in man since the fall of Adam.

But since you are not satisfied with my brief and obvious answer, you have only yourselves to blame for the consequences. You must now listen to a longer and more learned explanation, such as it is in my power to give.

As the church of science, however, debars the question "Why?" let us put the matter in a purely orthodox way: Man has two eyes, what *more* can he see with two than with one?

I will invite you to take a walk with me? We see
before us a wood. What is it that makes this real
wood contrast so favorably with a painted wood, no
matter how perfect the painting may be? What makes
the one so much more lovely than the other? Is it the
vividness of the coloring, the distribution of the lights

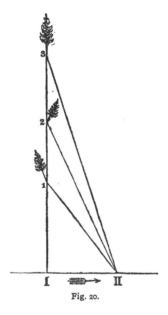

Fig. 20.

and the shadows? I think
not. On the contrary, it
seems to me that in this
respect painting can ac-
complish very much.

The cunning hand of
the painter can conjure up
with a few strokes of his
brush forms of wonderful
plasticity. By the help of
other means even more
can be attained. Photo-
graphs of reliefs are so
plastic that we often im-
agine we can actually lay
hold of the elevations and
depressions.

But one thing the painter never can give with the
vividness that nature does—the difference of near and
far. In the real woods you see plainly that you can
lay hold of some trees, but that others are inaccessibly
far. The picture of the painter is rigid. The picture
of the real woods changes on the slightest movement.

Now this branch is hidden behind that; now that behind this. The trees are alternately visible and invisible.

Let us look at this matter a little more closely. For convenience sake we shall remain upon the highway, I, II. (Fig. 20.) To the right and the left lies the forest. Standing at I, we see, let us say, three trees (1, 2, 3) in a line, so that the two remote ones are covered by the nearest. Moving further along, this changes. At II we shall not have to look round so far to see the remotest tree 3 as to see the nearer tree 2, nor so far to see this as to see 1. *Hence, as we move onward, objects that are near to us seem to lag behind as compared with objects that are remote from us, the lagging increasing with the proximity of the objects.* Very remote objects, towards which we must always look in the same direction as we proceed, appear to travel along with us.

If we should see, therefore, jutting above the brow of yonder hill the tops of two trees whose distance from us we were in doubt about, we should have in our hands a very easy means of deciding the question. We should take a few steps forward, say to the right, and the tree-top which receded most to the left would be the one nearer to us. In truth, from the amount of the recession a geometer could actually determine the distance of the trees from us without ever going near them. It is simply the scientific development of this perception that enables us to measure the distances of the stars.

Hence, from change of view in forward motion the distances of objects in our field of vision can be measured.

Rigorously, however, even forward motion is not necessary. For every observer is composed really of *two* observers. Man has *two* eyes. The right eye is a short step ahead of the left eye in the right-hand direction. Hence, the two eyes receive *different* pictures of the same woods. The right eye will see the near trees displaced to the left, and the left eye will see them displaced to the right, the displacement being greater, the greater the proximity. This difference is sufficient for forming ideas of distance.

We may now readily convince ourselves of the following facts :

1. With one eye, the other being shut, you have a very uncertain judgment of distances. You will find it, for example, no easy task, with one eye shut, to thrust a stick through a ring hung up before you; you will miss the ring in almost every instance.

2. You see the same object differently with the right eye from what you do with the left.

Place a lamp-shade on the table in front of you with its broad opening turned downwards, and look at it from above. (Fig. 21.) You will see with your right eye the image 2, with your left eye the image 1. Again, place the shade with its wide opening turned upwards; you will receive with your right eye the image 4, with your left eye the image 3. Euclid mentions phenomena of this character.

3. Finally, you know that it is easy to judge of distances with both eyes. Accordingly your judgment must spring in some way from a co-operation of the

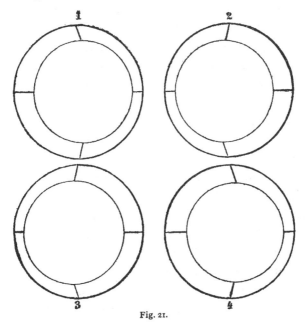

Fig. 21.

two eyes. In the preceding example the openings in the different images received by the two eyes seem displaced with respect to one another, and this displacement is sufficient for the inference that the one opening is nearer than the other.

I have no doubt that you, ladies, have frequently received delicate compliments upon your eyes, but I

feel sure that no one has ever told you, and I know not
whether it will flatter you, that you have in your eyes,
be they blue or black, little geometricians. You say
you know nothing of them? Well, for that matter,
neither do I. But the facts are as I tell you.

You understand little of geometry? I shall accept
that confession. Yet with the help of your two eyes
you judge of distances? Surely that is a geometrical
problem. And what is more, you know the solution
of this problem : for you estimate distances correctly.
If, then, *you* do not solve the problem, the little geometricians in your eyes must do it clandestinely and whis-
per the solution to you. I doubt not they are fleet little
fellows.

What amazes me most here is, that you know nothing about these little geometricians. But perhaps they
also know nothing about you. Perhaps they are models of punctuality, routine clerks who bother about
nothing but their fixed work. In that case we may
be able to deceive the gentlemen.

If we present to our right eye an image which looks
exactly like the lamp-shade for the right eye, and to
our left eye an image which looks exactly like a lamp-
shade for the left eye, we shall imagine that we see
the whole lamp-shade bodily before us.

You know the experiment. If you are practised in
squinting, you can perform it directly with the figure,
looking with your right eye at the right image, and
with your left eye at the left image. In this way the

experiment was first performed by Elliott. Improved and perfected, its form is Wheatstone's stereoscope, made so popular and useful by Brewster.

By taking two photographs of the same object from two different points, corresponding to the two eyes, a very clear three-dimensional picture of distant places or buildings can be produced by the stereoscope.

But the stereoscope accomplishes still more than this. It can visualise things for us which we never see with equal clearness in real objects. You know that if you move much while your photograph is being taken, your picture will come out like that of a Hindu deity, with several heads or several arms, which, at the spaces where they overlap, show forth with equal distinctness, so that we seem to see the one picture *through* the other. If a person moves quickly away from the camera before the impression is completed, the objects behind him will also be imprinted upon the photograph; the person will look transparent. Photographic ghosts are made in this way.

Some very useful applications may be made of this discovery. For example, if we photograph a machine stereoscopically, successively removing during the operation the single parts (where of course the impression suffers interruptions), we obtain a transparent view, endowed with all the marks of spatial solidity, in which is distinctly visualised the interaction of parts normally concealed. I have employed this method for

obtaining transparent stereoscopic views of anatomical structures.

You see, photography is making stupendous advances, and there is great danger that in time some
malicious artist will photograph his innocent patrons
with solid views of their most secret thoughts and
emotions.   How tranquil politics will then be ! What
rich harvests our detective force will reap !

*   *   *

By the joint action of the two eyes, therefore, we
arrive at our judgments of distances, as also of the
forms of bodies.

Permit me to mention here a few additional facts
connected with this subject, which will assist us in the
comprehension of certain phenomena in the history of
civilisation.

You have often heard, and know from personal experience, that remote objects appear perspectively
dwarfed.   In fact, it is easy to satisfy yourself that
you can cover the image of a man a few feet away
from you simply by holding up your finger a short distance in front of your eye.   Still, as a general rule,
you do not notice this shrinkage of objects.   On the
contrary, you imagine you see a man at the end of a
large hall, as large as you see him near by you.   For
your eye, in its measurement of the distances, makes
remote objects correspondingly larger.   The eye, so to
speak, is aware of this perspective contraction and is
not deceived by it, although its possessor is unconscious

of the fact. All persons who have attempted to draw from nature have vividly felt the difficulty which this superior dexterity of the eye causes the perspective conception. Not until one's judgment of distances is made uncertain, by their size, or from lack of points of reference, or from being too quickly changed, is the perspective rendered very prominent.

On sweeping round a curve on a rapidly moving railway train, where a wide prospect is suddenly opened up, the men upon distant hills appear like dolls.* You have at the moment, here, no known references for the measurement of distances. The stones at the entrance of a tunnel grow visibly larger as we ride towards it ; they shrink visibly in size as we ride from it.

Usually both eyes work together. As certain views are frequently repeated, and lead always to substantially the same judgments of distances, the eyes in time must acquire a special skill in geometrical constructions. In the end, undoubtedly, this skill is so increased that a single eye alone is often tempted to exercise that office.

Permit me to elucidate this point by an example. Is any sight more familiar to you than that of a vista down a long street? Who has not looked with hopeful

---

* This effect is particularly noticeable in the size of workmen on high chimneys and church-steeples—"steeple Jacks." When the cables were slung from the towers of the Brooklyn bridge (277 feet high), the men sent out in baskets to paint them, appeared, against the broad background of heaven and water, like flies.—*Trans.*

eyes time and again into a street and measured its
depth.   I will take you now into an art-gallery where
I will suppose you to see a picture representing a vista
into a street.   The artist has not spared his rulers to
get his perspective perfect.   The geometrician in your
left eye thinks, "Ah ha ! I have computed that case a
hundred times or more.   I know it by heart.   It is a
vista into a street," he continues ; " where the houses
are lower is the remote end."   The geometrician in
the right eye, too much at his ease to question his
possibly peevish comrade in the matter, answers the
same.   But the sense of duty of these punctual little
fellows is at once rearoused.   They set to work at their
calculations and immediately find that all the points
of the picture are equally distant from them, that is,
lie 'all upon a plane surface.

What opinion will you now accept, the first or the
second? If you accept the first you will see distinctly
the vista.   If you accept the second you will see nothing but a painted sheet of distorted images.

It seems to you a trifling matter to look at a picture and understand its perspective.   Yet centuries
elapsed before humanity came fully to appreciate this
trifle, and even the majority of you first learned it from
education.

I can remember very distinctly that at three years
of age all perspective drawings appeared to me as
gross caricatures of objects.   I could not understand
why artists made tables so broad at one end and so

narrow at the other. Real tables seemed to me just as broad at one end as at the other, because my eye made and interpreted its calculations without my intervention. But that the picture of the table on the plane surface was not to be conceived as a plane painted surface but stood for a table and so was to be imaged with all the attributes of extension was a joke that I did not understand. But I have the consolation that whole nations have not understood it.

Ingenuous people there are who take the mock murders of the stage for real murders, the dissembled actions of the players for real actions, and who can scarcely restrain themselves, when the characters of the play are sorely pressed, from running in deep indignation to their assistance. Others, again, can never forget that the beautiful landscapes of the stage are painted, that Richard III. is only the actor, Mr. Booth, whom they have met time and again at the clubs.

Both points of view are equally mistaken. To look at a drama or a picture properly one must understand that both are *shows*, simply *denoting* something real. A certain preponderance of the intellectual life over the sensuous life is requisite for such an achievement, where the intellectual elements are safe from destruction by the direct sensuous impressions. A certain liberty in choosing one's point of view is necessary, a sort of humor, I might say, which is strongly wanting in children and in childlike peoples.

Let us look at a few historical facts. I shall not

take you as far back as the stone age, although we
possess sketches from this epoch which show very original ideas of perspective.   But let us begin our sight-
seeing in the tombs and ruined temples of ancient
Egypt, where the numberless reliefs and gorgeous col-
orings have defied the ravages of thousands of years.

A rich and motley life is here opened to us.   We
find the Egyptians represented in all conditions of life.
What at once strikes our attention in these pictures
is the delicacy of their technical execution.   The con-
tours are extremely exact and distinct.   But on the
other hand only a few bright colors are found, un-
blended and without trace of transition. Shadows are
totally wanting.   The paint is laid on the surfaces in
equal thicknesses.

Shocking for the modern eye is the perspective.
All the figures are equally large, with the exception of
the king, whose form is unduly exaggerated. Near and
far appear equally large.   Perspective contraction is
nowhere employed.   A pond with water-fowl is represented flat, as if its surface were vertical.

Human figures are portrayed as they are never
seen, the legs from the side, the face in profile.   The
breast lies in its full breadth across the plane of representation.   The heads of cattle appear in profile,
while the horns lie in the plane of the drawing.   The
principle which the Egyptians followed might be best
expressed by saying that their figures are pressed in

the plane of the drawing as plants are pressed in a herbarium.

The matter is simply explained. If the Egyptians were accustomed to looking at things ingenuously with both eyes at once, the construction of perspective pictures in space could not be familiar to them. They saw all arms, all legs on real men in their natural lengths. The figures pressed into the planes resembled more closely, of course, in their eyes the originals than perspective pictures could.

This will be better understood if we reflect that painting was developed from relief. The minor dissimilarities between the pressed figures and the originals must gradually have compelled men to the adoption of perspective drawing. But physiologically the painting of the Egyptions is just as much justified as the drawings of our children are.

A slight advance beyond the Egyptians is shown by the Assyrians. The reliefs rescued from the ruined mounds of Nimrod at Mossul are, upon the whole, similar to the Egyptian reliefs. They were made known to us principally by Layard.

Painting enters on a new phase among the Chinese. This people have a marked feeling for perspective and correct shading, yet without being very logical in the application of their principles. Here, too, it seems, they took the first step but did not go far. In harmony with this immobility is their constitution, in which the muzzle and the bamboo-rod play sig-

nificant functions.   In accord with it, too, is their
language, which like the language of children has not
yet developed into a grammar, or, rather, according
to the modern conception, has not yet degenerated
into a grammar.   It is the same also with their music
which is satisfied with the five-toned scale.

The mural paintings at Herculaneum and Pompeii
are distinguished by grace of representation, as also
by a pronounced sense for perspective and correct illumination, yet they are not at all scrupulous in construction.   Here still we find abbreviations avoided.
But to offset this defect, the members of the body are
brought into unnatural positions, in which they appear
in their full lengths.   Abridgements are more frequently observed in clothed than in unclothed figures.

A satisfactory explanation of these phenomena first
occurred to me on the making of a few simple experiments which show how differently one may see the
same object, after some mastery of one's senses has

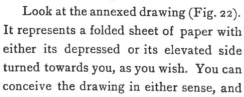 been attained, simply by the arbitrary
movement of the attention.

Look at the annexed drawing (Fig. 22).
It represents a folded sheet of paper with
either its depressed or its elevated side
turned towards you, as you wish. You can
conceive the drawing in either sense, and
in either case it will appear to you differently.

Fig. 22.

If, now, you have a real folded sheet of paper on
the table before you, with its sharp edges turned to-

wards you, you can, on looking at it with one eye, see the sheet alternately elevated, as it really is, or de‹ pressed. Here, however, a remarkable phenomenon is presented. When you see the sheet properly, neither illumination nor form presents anything conspicuous. When you see it bent back you see it perspectively distorted. Light and shadow appear much brighter or darker, or as if overlaid thickly with bright colors. Light and shadow now appear devoid of all cause. They no longer harmonise with the body's form, and are thus rendered much more prominent.

In common life we employ the perspective and illumination of objects to determine their forms and position. Hence we do not notice the lights, the shadows, and the distortions. They first powerfully enter consciousness when we employ a different construction from the usual spatial one. In looking at the planar image of a camera obscura we are amazed at the plenitude of the light and the profundity of the shadows, both of which we do not notice in real objects.

In my earliest youth the shadows and lights on pictures appeared to me as spots void of meaning. When I began to draw I regarded shading as a mere custom of artists. I once drew the portrait of our pastor, a friend of the family, and shaded, from no necessity, but simply from having seen something similar in other pictures, the whole half of his face black. I was subjected for this to a severe criticism on the part of my mother, and my deeply offended artist's pride is probably the reason that these facts remained so strongly impressed upon my memory.

You see, then, that many strange things, not only in the life of individuals, but also in that of humanity, and in the history of general civilisation, may be explained from the simple fact that man has two eyes.

Change man's eye and you change his conception of the world. We have observed the truth of this fact among our nearest kin, the Egyptians, the Chinese, and the lake-dwellers; how must it be among some of our remoter relatives,—with monkeys and other animals? Nature must appear totally different to animals equipped with substantially different eyes from those of men, as, for example, to insects. But for the present science must forego the pleasure of portraying this appearance, as we know very little as yet of the mode of operation of these organs.

It is an enigma even how nature appears to animals closely related to man; as to birds, who see scarcely anything with two eyes at once, but since their eyes are placed on opposite sides of their heads, have a separate field of vision for each.*

The soul of man is pent up in the prison-house of his head; it looks at nature through its two windows, the eyes. It would also fain know how nature looks through other windows. A desire apparently never to

---

* See Joh. Müller, *Vergleichende Physiologie des Gesichtssinnes*, Leipsic, 1826.

be fulfilled. But our love for nature is inventive, and here, too, much has been accomplished.

Placing before me an angular mirror, consisting of two plane mirrors slightly inclined to each other, I see my face twice reflected. In the right-hand mirror I obtain a view of the right side, and in the left-hand mirror a view of the left side, of my face. Also I shall see the face of a person standing in front

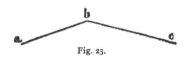

Fig. 23.

of me, more to the right with my right eye, more to the left with my left. But in order to obtain such widely different views of a face as those shown in the angular mirror, my two eyes would have to be set much further apart from each other than they actually are.

Squinting with my right eye at the image in the right-hand mirror, with my left eye at the image in the left-hand mirror, my vision will be the vision of a giant having an enormous head with his two eyes set far apart. This, also, is the impression which my own face makes upon me. I see it now, single and solid. Fixing my gaze, the relief from second to second is magnified, the eyebrows start forth prominently from above the eyes, the nose seems to grow a foot in length, my mustache shoots forth like a fountain from my lip, the teeth seem to retreat immeasurably. But by far the most horrible aspect of the phenomenon is the nose.

Interesting in this connexion is the telestereoscope

of Helmholtz. In the telestereoscope we view a land-scape by looking with our right eye (Fig. 24) through the mirror *a* into the mirror *A*, and with our left eye through the mirror *b* into the mirror *B*. The mirrors *A* and *B* stand far apart. Again we see with the widely separated eyes of a giant. Everything appears dwarfed and near us. The distant mountains look like

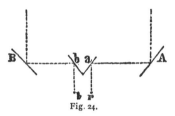

Fig. 24.

moss-covered stones at our feet. Between, you see the reduced model of a city, a veritable Liliput. You are tempted almost to stroke with your hand the soft forest and city, did you not fear that you might prick your fingers on the sharp, needle-shaped steeples, or that they might crackle and break off.

Liliput is no fable. We need only Swift's eyes, the telestereoscope, to see it.

Picture to yourself the reverse case. Let us sup-pose ourselves so small that we could take long walks in a forest of moss, and that our eyes were correspond-ingly near each other. The moss-fibres would appear like trees. On them we should see strange, unshapely monsters creeping about. Branches of the oak-tree, at whose base our moss-forest lay, would seem to us dark, immovable, myriad-branched clouds, painted high on the vault of heaven; just as the inhabitants of Saturn, forsooth, might see their enormous ring.

On the tree-trunks of our mossy woodland we should
find colossal globes several feet in diameter, brilliantly
transparent, swayed by the winds with slow, peculiar
motions. We should approach inquisitively and should
find that these globes, in which here and there animals were gaily sporting, were liquid globes, in fact
that they were water. A short, incautious step, the
slightest contact, and woe betide us, our arm is irresist-
ibly drawn by an invisible power into the interior of
the sphere and held there unrelentingly fast! A drop
of dew has engulfed in its capillary maw a manikin,
in revenge for the thousands of drops that its big human counterparts have quaffed at breakfast. Thou
shouldst have known, thou pygmy natural scientist,
that with thy present puny bulk thou shouldst not joke
with capillarity!

My terror at the accident brings me back to my
senses. I see I have turned idyllic. You must pardon
me. A patch of greensward, a moss or heather forest
with its tiny inhabitants have incomparably more
charms for me than many a bit of literature with its
apotheosis of human character. If I had the gift of
writing novels I should certainly not make John and
Mary my characters. Nor should I transfer my loving
pair to the Nile, nor to the age of the old Egyptian
Pharaohs, although perhaps I should choose that time
in preference to the present. For I must candidly
confess that I hate the rubbish of history, interesting
though it may be as a mere phenomenon, because we

cannot simply observe it but must also *feel* it, because
it comes to us mostly with supercilious arrogance,
mostly unvanquished. The hero of my novel would be
a cockchafer, venturing forth in his fifth year for the
first time with his newly grown wings into the light,
free air. Truly it could do no harm if man would thus
throw off his inherited and acquired narrowness of
mind by making himself acquainted with the world-
view of allied creatures. He could not help gaining
incomparably more in this way than the inhabitant of
a small town would in circumnavigating the globe and
getting acquainted with the views of strange peoples.

*   *   *

I have now conducted you, by many paths and by-
ways, rapidly over hedge and ditch, to show you what
wide vistas we may reach in every field by the rigor-
ous pursuit of a single scientific fact. A close examination of the two eyes of man has conducted us not
only into the dim recesses of humanity's childhood,
but has also carried us far beyond the bourne of human
life.

It has surely often struck you as strange that the
sciences are divided into two great groups , that the
so-called humanistic sciences, belonging to the so-
called "higher education," are placed in almost a hos-
tile attitude to the natural sciences.

I must confess I do not overmuch believe in this
partition of the sciences. I believe that this view will
appear as childlike and ingenuous to a matured age

as the want of perspective in the old paintings of Egypt
do to us. Can it really be that "higher culture" is only
to be obtained from a few old pots and palimpsests,
which are at best mere scraps of nature, or that more
is to be learned from them alone than from all the rest
of nature? I believe that both these sciences are simply parts of the same science, which have begun at
different ends. If these two ends still act towards
each other as the Montagues and Capulets, if their re-
tainers still indulge in lively tilts, I believe that after
all they are not in earnest. On the one side there is
surely a Romeo, and on the other a Juliet, who, some
day, it is hoped, will unite the two houses with a less
tragic sequel than that of the play.

Philology began with the unqualified reverence and
apotheosis of the Greeks. Now it has begun to draw
other languages, other peoples and their histories, into
its sphere; it has, through the mediation of comparative linguistics, already struck up, though as yet somewhat cautiously, a friendship with physiology.

Physical science began in the witch's kitchen. It
now embraces the organic and inorganic worlds, and
with the physiology of articulation and the theory of
the senses, has even pushed its researches, at times
impertinently, into the province of mental phenomena.

In short, we come to the understanding of much
within us solely by directing our glance without, and
vice versa. Every object belongs to both sciences.
You, ladies, are very interesting and difficult problems

for the psychologist, but you are also extremely pretty
phenomena of nature.   Church and State are objects
of the historian's research, but not less phenomena of
nature, and in part, indeed, very curious phenomena.
If the historical sciences have inaugurated wide ex-
tensions of view by presenting to us the thoughts of
new and strange peoples, the physical sciences in a
certain sense do this in a still greater degree.   In
making man disappear in the All, in annihilating him,
so to speak, they force him to take an unprejudiced
position without himself, and to form his judgments by
a different standard than that of the petty human.

But if you should ask me now why man has two
eyes, I should answer :

That he may look at nature justly and accurately;
that he may come to understand that he himself, with
all his views, correct and incorrect, with all his *haute
politique,* is simply an evanescent shred of nature ;
that, to speak with Mephistopheles, he is a part of the
part, and that it is absolutely unjustified,

> " For man, the microcosmic fool, to see
> Himself a whole so frequently."

ON SYMMETRY.*

AN ancient philosopher once remarked that people who cudgelled their brains about the nature of the moon reminded him of men who discussed the laws and institutions of a distant city of which they had heard no more than the name. The true philosopher, he said, should turn his glance within, should study himself and his notions of right and wrong; only thence could he derive real profit.

This ancient formula for happiness might be restated in the familiar words of the Psalm:

"Dwell in the land, and verily thou shalt be fed."

To-day, if he could rise from the dead and walk about among us, this philosopher would marvel much at the different turn which matters have taken.

* Delivered before the German Casino of Prague, in the winter of 1871.
A fuller treatment of the problems of this lecture will be found in my *Contributions to the Analysis of the Sensations* (Jena, 1886), English Translation, Chicago, 1895. J. P. Soret, *Sur la perception du beau* (Geneva, 1892), also regards repetition as a principle of æsthetics. His discussions of the *æsthetical* side of the subject are much more detailed than mine. But with respect to the psychological and physiological foundation of the principle, I am convinced that the *Contributions to the Analysis of the Sensations* go deeper.— MACH (1894).

The motions of the moon and the other heavenly
bodies are accurately known.   Our knowledge of the
motions of our own body is by far not so complete.
The mountains and natural divisions of the moon have
been accurately outlined on maps, but physiologists
are just beginning to find their way in the geography
of the brain. The chemical constitution of many fixed
stars has already been investigated.   The chemical
processes of the animal body are questions of much
greater difficulty and complexity.   We have our *Mécanique celeste.*   But a *Mécanique sociale* or a *Mécanique
morale* of equal trustworthiness remains to be written.

Our philosopher would indeed admit that we have
made great progress.   But we have not followed his
advice.   The patient has recovered, but he took for his
recovery exactly the opposite of what the doctor pre-
scribed.

Humanity is now returned, much wiser, from its
journey in celestial space, against which it was so
solemnly warned. Men, after having become acquainted
with the great and simple facts of the world without,
are now beginning to examine critically the world
within. It sounds absurd, but it is true, that only after
we have thought about the moon are we able to take
up ourselves. It was necessary that we should acquire
simple and clear ideas in a less complicated domain,
before we entered the more intricate one of psychology, and with these ideas astronomy principally furnished us.

To attempt any description of that stupendous movement, which, originally springing out of the physical sciences, went beyond the domain of physics and is now occupied with the problems of psychology, would be presumptuous in this place. I shall only attempt here, to illustrate to you by a few simple examples the methods by which the province of psychology can be reached from the facts of the physical world—especially the adjacent province of sense-perception. And I wish it to be remembered that my brief attempt is not to be taken as a measure of the present state of such scientific questions.

*   *   *

It is a well-known fact that some objects please us, while others do not. Generally speaking, anything that is constructed according to fixed and logically followed rules, is a product of tolerable beauty. We see thus nature herself, who always acts according to fixed rules, constantly producing such pretty things. Every day the physicist is confronted in his workshop with the most beautiful vibration-figures, tone-figures, phenomena of polarisation, and forms of diffraction.

A rule always presupposes a repetition. Repetitions, therefore, will probably be found to play some important part in the production of agreeable effects. Of course, the nature of agreeable effects is not exhausted by this. Furthermore, the repetition of a physical event becomes the source of agreeable effects

only when it is connected with a repetition of sensations.

An excellent example that repetition of sensations
is a source of agreeable effects is furnished by the
copy-book of every schoolboy, which is usually a treas-
ure-house of such things, and only in need of an Abbé
Domenech to become celebrated. Any figure, no matter how crude or poor, if several times repeated, with
the repetitions placed in line, will produce a tolerable
frieze.

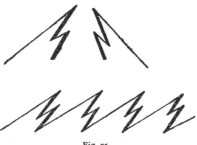

Fig. 25.

Also the pleasant effect of symmetry is due to the
repetition of sensations. Let us abandon ourselves a
moment to this thought, yet not imagine when we have
developed it, that we have fully exhausted the nature
of the agreeable, much less of the beautiful.

First, let us get a clear conception of what symmetry is. And in preference to a definition let us take
a living picture. You know that the reflexion of an
object in a mirror has a great likeness to the object itself. All its proportions and outlines are the same.

Yet there is a difference between the object and its reflexion in the mirror, which you will readily observe. Hold your right hand before a mirror, and you will see in the mirror a left hand. Your right glove will produce its mate in the glass. For you could never use the reflexion of your right glove, if it were present to you as a real thing, for covering your right hand, but only for covering your left. Similarly, your right ear will give as its reflexion a left ear ; and you will at once perceive that the left half of your body could very easily be substituted for the reflexion of your right half. Now just as in the place of a missing right ear a left ear cannot be put, unless the lobule of the ear be turned upwards, or the opening into the concha backwards, so, despite all similarity of form, the reflexion of an object can never take the place of the object itself.*

The reason of this difference between the object and its reflexion is simple. The reflexion appears as far behind the mirror as the object is in front of it. The parts of the object, accordingly, which are nearest the mirror will also be nearest the mirror in the reflexion. Consequently, the succession of the parts in the reflexion will be reversed, as may best be seen in the reflexion of the face of a watch or of a manuscript.

It will also be readily seen, that if a point of the object be joined with its reflexion in the image, the line of junction will cut the mirror at right angles and be

---

* Kant, in his *Prolegomena zu jeder künftigen Metaphysik*, also refers to this fact, but for a different purpose.

bisected by it. This holds true of all corresponding points of object and image.

If, now, we can divide an object by a plane into two halves so that each half, as seen in the reflecting plane of division, is a reproduction of the other half, such an object is termed symmetrical, and the plane of division is called the plane of symmetry.

If the plane of symmetry is vertical, we can say that the body is vertically symmetrical. An example of vertical symmetry is a Gothic cathedral.

If the plane of symmetry is horizontal, we can say that the object is horizontally symmetrical. A landscape on the shores of a lake with its reflexion in the water, is a system of horizontal symmetry.

Exactly here is a noticeable difference. The vertical symmetry of a Gothic cathedral strikes us at once, whereas we can travel up and down the whole length of the Rhine or the Hudson without becoming aware of the symmetry between objects and their reflexions in the water. Vertical symmetry pleases us, whilst horizontal symmetry is indifferent, and is noticed only by the experienced eye.

Whence arises this difference? I say from the fact that vertical symmetry produces a repetition of the same sensation, while horizontal symmetry does not. I shall now show that this is so.

Let us look at the following letters :

d   b

q   p

It is a fact known to all mothers and teachers, that children in their first attempts to read and write, constantly confound d and b, and q and p, but never d and q, or b and p.   Now d and b and q and p are the two halves of a *vertically* symmetrical figure, while d and q, and b and p are two halves of a *horizontally* symmetrical figure.   The first two are confounded; but confusion is only possible of things that excite in us the same or similar sensations.

Figures of two flower-girls are frequently seen on the decorations of gardens and of drawing-rooms, one of whom carries a flower-basket in her right hand and the other a flower-basket in her left.   All know how apt we are, unless we are very careful, to confound these figures with one another.

While turning a thing round from right to left is scarcely noticed, the eye is not at all indifferent to the turning of a thing upside down.   A human face which has been turned upside down is scarcely recognisable as a face, and makes an impression which is altogether strange.   The reason of this is not to be sought in the unwontedness of the sight, for it is just as difficult to recognise an arabesque that has been inverted, where there can be no question of a habit.   This curious fact is the foundation of the familiar jokes played with the portraits of unpopular personages, which are so drawn that in the upright position of the page an exact picture of the person is presented, but on being inverted some popular animal is shown.

It is a fact, then, that the two halves of a vertically symmetrical figure are easily confounded and that they therefore probably produce very nearly the same sensations. The question, accordingly, arises, *why* do the two halves of a vertically symmetrical figure produce the same or similar sensations? The answer is: Because our apparatus of vision, which consists of our eyes and of the accompanying muscular apparatus is itself vertically symmetrical.*

Whatever external resemblances one eye may have with another they are still not alike. The right eye of a man cannot take the place of a left eye any more than a left ear or left hand can take the place of a right one. By artificial means, we can change the part which each of our eyes plays. (Wheatstone's pseudo-scope.) But we then find ourselves in an entirely new and strange world. What is convex appears concave; what is concave, convex. What is distant appears near, and what is near appears far.

The left eye is the reflexion of the right. And the light-feeling retina of the left eye is a reflexion of the light-feeling retina of the right, in all its functions.

The lense of the eye, like a magic lantern, casts images of objects on the retina. And you may picture to yourself the light-feeling retina of the eye, with its countless nerves, as a hand with innumerable fingers, adapted to feeling light. The ends of the visual nerves, like our fingers, are endowed with varying degrees of

* Compare Mach, *Fichte's Zeitschrift für Philosophie*, 1864, p. 1.

sensitiveness. The two retinæ act like a right and a left hand ; the sensation of touch and the sensation of light in the two instances are similar.

Examine the right-hand portion of this letter T : namely, T. Instead of the two retinæ on which this image falls, imagine feeling the object, my two hands. The Γ, grasped with the right hand, gives a different sensation from that which it gives when grasped with the left. But if we turn our character about from right to left, thus : ꓶ, it will give the same sensation in the left hand that it gave before in the right. The sensation is repeated.

If we take a whole T, the right half will produce in the right hand the same sensation that the left half produces in the left, and *vice versa.*

The symmetrical figure gives the same sensation twice.

If we turn the T over thus : ⊢ , or invert the half T thus : L, so long as we do not change the position of our hands we can make no use of the foregoing reasoning.

The retinæ, in fact, are exactly like our two hands. They, too, have their thumbs and index fingers, though they are thousands in number ; and we may say the thumbs are on the side of the eye near the nose, and the remaining fingers on the side away from the nose.

With this I hope to have made perfectly clear that the pleasing effect of symmetry is chiefly due to the repetition of sensations, and that the effect in question takes place in symmetrical figures, only where there is a repetition of sensation. The pleasing effect of regular figures, the preference which straight lines, especially vertical and horizontal straight lines, enjoy, is founded on a similar reason. A straight line, both in a horizontal and in a vertical position, can cast on the two retinæ the same image, which falls moreover on symmetrically corresponding spots. This also, it would appear, is the reason of our psychological preference of straight to curved lines, and not their property of being the shortest distance between two points. The straight line is felt, to put the matter briefly, as symmetrical to itself, which is the case also with the plane. Curved lines are felt as deviations from straight lines, that is, as deviations from symmetry.* The presence of a sense for symmetry in people possessing only one eye from birth, is indeed a riddle. Of course, the sense of symmetry, although primarily acquired by means of the eyes, cannot be wholly limited to the visual organs. It must also be deeply rooted in other parts of the organism by ages of practice and can thus not be eliminated forthwith by the loss of one eye. Also, when an eye is lost, the sym-

---

*The fact that the first and second differential coefficients of a curve are directly seen, but the higher coefficients not, is very simply explained. The first gives the position of the tangent, the declination of the straight line from the position of symmetry, the second the declination of the curve from the straight line. It is, perhaps, not unprofitable to remark here that the ordinary method of testing rulers and plane surfaces (by reversed applications) ascertains the deviation of the object from symmetry to itself.

metrical muscular apparatus is left, as is also the symmetrical apparatus of innervation.

It appears, however, unquestionable that the phenomena mentioned have, in the main, their origin in the peculiar structure of our eyes. It will therefore be seen at once that our notions of what is beautiful and ugly would undergo a change if our eyes were different. Also, if this view is correct, the theory of the so-called eternally beautiful is somewhat mistaken. It can scarcely be doubted that our culture, or form of civilisation, which stamps upon the human body its unmistakable traces, should not also modify our conceptions of the beautiful. Was not formerly the development of all musical beauty restricted to the narrow limits of a five-toned scale ?

The fact that a repetition of sensations is productive of pleasant effects is not restricted to the realm of the visible. To-day, both the musician and the physicist know that the harmonic or the melodic addition of one tone to another affects us agreeably only when the added tone reproduces a part of the sensation which the first one excited. When I add an octave to a fundamental tone, I hear in the octave a part of what was heard in the fundamental tone. (Helmholtz.) But it is not my purpose to develop this idea fully here.* We shall only ask to-day, whether there is anything similar to the symmetry of figures in the province of sounds.

* See the lecture *On the Causes of Harmony.*

Look at the reflexion of your piano in the mirror. You will at once remark that you have never seen such a piano in the actual world, for it has its high keys to the left and its low ones to the right. Such pianos are not manufactured.

If you could sit down at such a piano and play in your usual manner, plainly every step which you imagined you were performing in the upward scale would be executed as a corresponding step in the downward scale. The effect would be not a little surprising.

For the practised musician who is always accustomed to hearing certain sounds produced when certain keys are struck, it is' quite an anomalous spectacle to watch a player in the glass and to observe that he always does the opposite of what we hear.

But still more remarkable would be the effect of attempting to strike a harmony on such a piano. For a melody it is not indifferent whether we execute a step in an upward or a downward scale. But for a harmony, so great a difference is not produced by reversal. I always retain the same consonance whether I add to a fundamental note an upper or a lower third. Only the order of the intervals of the harmony is reversed. In point of fact, when we execute a movement in a major key on our reflected piano, we hear a sound in a minor key, and *vice versa*.

It now remains to execute the experiments indicated. Instead of playing upon the piano in the mirror, which is impossible, or of having a piano of this kind built, which would be somewhat expensive, we may perform our experiments in a simpler manner, as follows:

1) We play on our own piano in our usual manner, look into the mirror, and then repeat on our real piano what we see in the mirror. In this way we transform all steps upwards into corresponding steps downwards. We play a movement, and then another movement, which, with respect to the key-board, is symmetrical to the first.

2) We place a mirror beneath the music in which the notes are reflected as in a body of water, and play according to the notes in the mirror. In this way also, all steps upwards are changed into corresponding, equal steps downwards.

3) We turn the music upside down and read the notes from right to left and from below upwards. In doing this, we must regard all sharps as flats and all flats as sharps, because they correspond to half lines and spaces. Besides, in this use of the music we can only employ the bass clef, as only in this clef are the notes not changed by symmetrical reversal.

You can judge of the effect of these experiments from the examples which appear in the annexed musical cut. (Page 102.) The movement which appears in the upper lines is symmetrically reversed in the lower.

The effect of the experiments may be briefly formulated. The melody is rendered unrecognisable. The

Fig. 26.

(See pages 101 and 103.)

harmony suffers a transposition from a major into a minor key and *vice versa*. The study of these pretty effects, which have long been familiar to physicists and musicians, was revived some years ago by Von Oettingen.*

Now, although in all the preceding examples I have transposed steps upward into equal and similar steps downward, that is, as we may justly say, have played for every movement the movement which is symmetrical to it, yet the ear notices either little or nothing of symmetry. The transposition from a major to a minor key is the sole indication of symmetry remaining. The symmetry is there for the mind, but is wanting for sensation. No symmetry exists for the ear, because a reversal of musical sounds conditions no repetition of sensations. If we had an ear for height and an ear for depth, just as we have an eye for the right and an eye for the left, we should also find that symmetrical sound-structures existed for our auditory organs. The contrast of major and minor for the ear corresponds to inversion for the eye, which is also only symmetry for the mind, but not for sensation.

By way of supplement to what I have said, I will add a brief remark for my mathematical readers.

Our musical notation is essentially a graphical representation of a piece of music in the form of curves, where the time is the abscissæ, and the logarithms of

* A. von Oettingen, *Harmoniesystem in dualer Entwicklung.* Leipsic and Dorpat, 1866.

the number of vibrations the ordinates. The deviations of musical notation from this principle are only such as facilitate interpretation, or are due to historical accidents.

If, now, it be further observed that the sensation of pitch also is proportional to the logarithm of the number of vibrations, and that the intervals between the notes correspond to the differences of the logarithms of the numbers of vibrations, the justification will be found in these facts of calling the harmonies and melodies which appear in the mirror, symmetrical to the original ones.

*   *   *

I simply wish to bring home to your minds by these fragmentary remarks that the progress of the physical sciences has been of great help to those branches of psychology that have not scorned to consider the results of physical research. On the other hand, psychology is beginning to return, as it were, in a spirit of thankfulness, the powerful stimulus which it received from physics.

The theories of physics which reduce all phenomena to the motion and equilibrium of smallest particles, the so-called molecular theories, have been gravely threatened by the progress of the theory of the senses and of space, and we may say that their days are numbered.

I have shown elsewhere* that the musical scale is

* Compare Mach's *Zur Theorie des Gehörorgans*, Vienna Academy, 1863.

simply a species of space—a space, however, of only one dimension, and that, a one-sided one. If, now, a person who could only hear, should attempt to develop a conception of the world in this, his linear space, he would become involved in many difficulties, as his space would be incompetent to comprehend the many sides of the relations of reality. But is it any more justifiable for us, to attempt to force the whole world into the space of our eye, in aspects in which it is not accessible to the eye? Yet this is the dilemma of all molecular theories.

We possess, however, a sense, which, with respect to the scope of the relations which it can comprehend, is richer than any other. It is our reason. This stands above the senses. It alone is competent to found a permanent and sufficient view of the world. The mechanical conception of the world has performed wonders since Galileo's time. But it must now yield to a broader view of things. A further development of this idea is beyond the limits of my present purpose.

One more point and I have done. The advice of our philosopher to restrict ourselves to what is near at hand and useful in our researches, which finds a kind of exemplification in the present cry of inquirers for limitation and division of labor, must not be too slavishly followed. In the seclusion of our closets, we often rack our brains in vain to fulfil a work, the means of accomplishing which lies before our very doors. If the inquirer must be perforce a shoemaker, tapping constantly at his last, it may perhaps be permitted him to be a shoemaker of the type of Hans Sachs, who did not deem it beneath him to take a look now and then at his neighbor's work and to comment on the latter's doings.

Let this be my apology, therefore, if I have forsaken for a moment to-day the last of my specialty.

ON THE FUNDAMENTAL CONCEPTS
OF ELECTROSTATICS.*

THE task has been assigned me to develop before you in a popular manner the fundamental quantitative concepts of electrostatics—"quantity of electricity," "potential," "capacity," and so forth. It would not be difficult, even within the brief limits of an hour, to delight the eye with hosts of beautiful experiments and to fill the imagination with numerous and varied conceptions. But we should, in such a case, be still far from a lucid and easy grasp of the phenomena. The means would still fail us for reproducing the facts accurately in thought—a procedure which for the theoretical and practical man is of equal importance. These means are the *metrical concepts* of electricity.

As long as the pursuit of the facts of a given province of phenomena is in the hands of a few isolated investigators, as long as every experiment can be easily repeated, the fixing of the collected facts by provisional

* A lecture delivered at the International Electrical Exhibition, in Vienna, on September 4, 1883.

description is ordinarily sufficient. But the case is different when the whole world must make use of the results reached by many, as happens when the science acquires broader foundations and scope, and particularly so when it begins to supply intellectual nourishment to an important branch of the practical arts, and to draw from that province in return stupendous empirical results. Then the facts must be so described that individuals in all places and at all times can, from a few easily obtained elements, put the facts accurately together in thought, and reproduce them from the description. This is done with the help of the metrical concepts and the international measures.

The work which was begun in this direction in the period of the purely scientific development of the science, especially by Coulomb (1784), Gauss (1833), and Weber (1846), was powerfully stimulated by the requirements of the great technical undertakings manifested since the laying of the first transatlantic cable, and brought to a brilliant conclusion by the labors of the British Association, 1861, and of the Paris Congress, 1881, chiefly through the exertions of Sir William Thomson.

It is plain, that in the time allotted to me I cannot conduct you over all the long and tortuous paths which the science has actually pursued, that it will not be possible at every step to remind you of all the little precautions for the avoidance of error which the early steps have taught us. On the contrary, I must make

shift with the simplest and rudest tools.   I shall conduct you by the shortest paths from the facts to the
ideas, in doing which, of course, it will not be possible
to anticipate all the stray and chance ideas which may
and must arise from prospects into the by-paths which
we leave untrodden.

*   *   *

Here are two small, light bodies (Fig. 27) of equal
size, freely suspended, which we "electrify" either

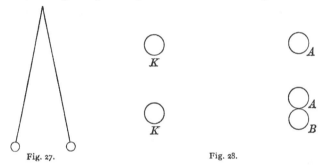

Fig. 27.   Fig. 28.

by friction with a third body or by contact with a body
already electrified.   At once a repulsive force is set
up which drives the two bodies away from each other
in opposition to the action of gravity.   This force could
accomplish anew the same mechanical work which
was expended to produce it.*

Coulomb, now, by means of delicate experiments
with the torsion-balance, satisfied himself that if the
bodies in question, say at a distance of two centimetres, repélled each other with the same force with

---

* If the two bodies were oppositely electrified they would exert attractions
upon each other.

which a milligramme-weight strives to fall to the ground, at half that distance, or at one centimetre, they would repel each other with the force of four milligrammes, and at double that distance, or at four centimetres, they would repel each other with the force of only one-fourth of a milligramme. He found that the electrical force acts inversely as the square of the distance.

Let us imagine, now, that we possessed some means of measuring electrical repulsion by weights, a means which would be supplied, for example, by our electrical pendulums; then we could make the following observation.

The body *A* (Fig. 28) is repelled by the body *K* at a distance of two centimetres with a force of one milligramme. If we touch *A*, now, with an equal body *B*, the half of this force of repulsion will pass to the body *B*; both *A* and *B*, now, at a distance of two centimetres from *K*, are repelled only with the force of one-half a milligramme. But both together are repelled still with the force of one milligramme. Hence, *the divisibility of electrical force* among bodies in contact *is a fact*. It is a useful, but by no means a necessary supplement to this fact, to imagine an electrical fluid present in the body *A*, with the quantity of which the electrical force varies, and half of which flows over to *B*. For, in the place of the new physical picture, thus, an old, familiar one is substituted, which moves spontaneously in its wonted courses.

Adhering to this idea, we define the *unit* of electrical quantity, according to the now almost universally adopted centimetre-gramme-second (C. G. S.) system, as that quantity which at a distance of one centimetre repels an equal quantity with unit of force, that is, with a force which in one second would impart to a mass of one gramme a velocity-increment of a centimetre. As a gramme mass acquires through the action of gravity a velocity-increment of about 981 centimetres in a second, accordingly, a gramme is attracted to the earth with 981, or, in round numbers, 1000 units of force of the centimetre-gramme-second system, while a milligramme-weight would strive to fall to the earth with approximately the unit force of this system.

We may easily obtain by this means a clear idea of what the unit quantity of electricity is. Two small bodies, *K*, weighing each a gramme, are hung up by vertical threads, five metres in length and almost weightless, so as to touch each other. If the two bodies be equally electrified and move apart upon electrification to a distance of one centimetre, their charge is approximately equivalent to the electrostatic unit of electric quantity, for the repulsion then holds in equilibrium a gravitational force-component of approximately one milligramme, which strives to bring the bodies together.

Vertically beneath a small sphere suspended from the equilibrated beam of a balance a second sphere is placed at a distance of a centimetre. If both be equally electrified the sphere suspended from the balance will be rendered apparently lighter by the repulsion. If by adding a weight of one milligramme equilibrium is restored, each of the spheres contains in round numbers the electrostatic unit of electrical quantity.

In view of the fact that the same electrical bodies exert at different distances different forces upon one another, exception might be taken to the measure of quantity here developed. What kind of a quantity is that which now weighs more, and now weighs less, so to speak? But this apparent deviation from the method of determination commonly used in practical life, that by weight, is, closely considered, an agreement. On a high mountain a heavy mass also is less powerfully attracted to the earth than at the level of the sea, and if it is permitted us in our determinations to neglect the consideration of level, it is only because the comparison of a body with fixed conventional weights is invariably effected at the same level. In fact, if we were to make one of the two weights equilibrated on our balance approach sensibly to the centre of the earth, by suspending it from a very long thread, as Prof. von Jolly of Munich suggested, we should make the gravity of that weight, its heaviness, proportionately greater.

Let us picture to ourselves, now, two different electrical fluids, a positive and a negative fluid, of such nature that the particles of the one attract the particles of the other according to the law of the inverse squares, but the particles of the same fluid repel each other by the same law ; in non-electrical bodies let us imagine the two fluids uniformly distributed in equal quantities, in electric bodies one of the two in excess; in conductors, further, let us imagine the fluids mobile, in non-conductors immobile ; having formed such pictures, we possess the conception which Coulomb developed and to which he gave mathematical precision. We have only to give this conception free play in our minds and we shall see as in a clear picture the fluid particles, say of a positively charged conductor, receding from one another as far as they can, all making for the surface of the conductor and there seeking out the prominent parts and points until the greatest possible amount of work has been performed. On increasing the size of the surface, we see a dispersion, on decreasing its size we see a condensation of the particles. In a second, non-electrified conductor brought into the vicinity of the first, we see the two fluids immediately separate, the positive collecting itself on the remote and the negative on the adjacent side of its surface. In the fact that this conception reproduces, lucidly and spontaneously, all the data which arduous research only slowly and gradually discovered, is contained its advantage and scientific value. With this, too, its value is exhausted. We must not seek in nature for the two hypothetical fluids which we have added as simple mental adjuncts, if we would not go astray. Coulomb's view may be replaced by a totally different one, for example, by that of Faraday, and the most proper course is always, after the general survey is obtained, to go back to the actual facts, to the electrical forces.

We will now make ourselves familiar with the concept of electrical quantity, and with the method of measuring or estimating it. Imagine a common Leyden jar (Fig. 29), the inner and outer coatings of which are connected together by means of two common me-

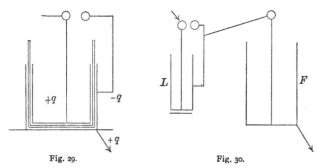

Fig. 29.   Fig. 30.

tallic knobs placed about a centimetre apart. If the inside coating be charged with the quantity of electricity $+q$, on the outer coating a distribution of the electricities will take place. A positive quantity almost equal* to the quantity $+q$ flows off to the earth, while a corresponding quantity $-q$ is still left on the outer coating. The knobs of the jar receive their portion of these quantities and when the quantity q is sufficiently great a rupture of the insulating air between the knobs,

---

*The quantity which flows off is in point of fact less than q. It would be equal to the quantity q only if the inner coating of the jar were wholly encompassed by the outer coating.

accompanied by the self-discharge of the jar, takes place. For any given distance and size of the knobs, a charge of a definite electric quantity q is always necessary for the spontaneous discharge of the jar.

Let us insulate, now, the outer coating of a Lane's unit jar L, the jar just described, and put in connexion with it the inner coating of a jar F exteriorly connected with the earth (Fig. 30). Every time that L is charged with $+q$, a like quantity $+q$ is collected on the inner coating of F, and the spontaneous discharge of the jar L, which is now again empty, takes place. The number of the discharges of the jar L furnishes us, thus, with a measure of the quantity collected in the jar F, and if after 1, 2, 3, . . . spontaneous discharges of L the jar F is

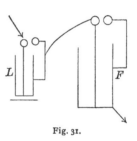

Fig. 31.

discharged, it is evident that the charge of F has been proportionately augmented.

Let us supply now, to effect the spontaneous discharge, the jar F with knobs of the same size and at the same distance apart as those of the jar L (Fig. 31). If we find, then, that five discharges of the unit jar take place before one spontaneous discharge of the jar F occurs, plainly the jar F, for equal distances between the knobs of the two jars, equal striking distances, is able to hold five times the quantity of electricity that L can, that is, has five times the *capacity* of L.*

We will now replace the unit jar L, with which we measure electricity, so to speak, *into* the jar F, by a Franklin's pane, consisting of two parallel flat metal plates (Fig. 32), separated only by air. If here, for example, thirty spontaneous discharges of the pane are sufficient to fill the jar, ten discharges will be found

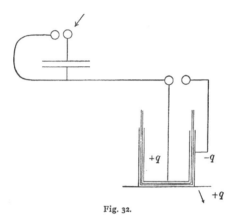

Fig. 32.

sufficient if the air-space between the two plates be filled with a cake of sulphur. Hence, the capacity of a Franklin's pane of sulphur is about three times greater than that of one of the same shape and size

---

*Rigorously, of course, this is not correct. First, it is to be noted that the jar L is discharged simultaneously with the electrode of the machine. The jar F, on the other hand, is always discharged simultaneously with the outer coating of the jar L. Hence, if we call the capacity of the electrode of the machine E, that of the unit jar L, that of the outer coating of L, A, and that of the principal jar F, then this equation would exist for the example in the text: $(F+A)/(L+E) = 5$. A cause of further departure from absolute exactness is the residual charge.

made of air, or, as it is the custom to say, the specific inductive capacity of sulphur (that of air being taken as the unit) is about 3.* We are here arrived at a very simple fact, which clearly shows us the significance of the number called *dielectric constant*, or *specific inductive capacity*, the knowledge of which is so important for the theory of submarine cables.

Let us consider a jar *A*, which is charged with a certain quantity of electricity. We can discharge the jar directly. But we can also discharge the jar *A*

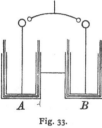

A	B

Fig. 33.   Fig. 34.

(Fig. 33) partly into a jar *B*, by connecting the two outer coatings with each other. In this operation a portion of the quantity of electricity passes, accompanied by sparks, into the jar *B*, and we now find both jars charged.

It may be shown as follows that the conception of

* Making allowance for the corrections indicated in the preceding footnote, I have obtained for the dielectric constant of sulphur the number 3.2, which agrees practically with the results obtained by more delicate methods. For the highest attainable precision one should by rights immerse the two plates of the condenser first wholly in air and then wholly in sulphur, if the ratio of the capacities is to correspond to the dielectric constant. In point of fact, however, the error which arises from inserting simply a plate of sulphur that exactly fills the space between the two plates, is of no consequence.

a constant quantity of electricity can be regarded as
the expression of a pure fact.   Picture to yourself any
sort of electrical conductor (Fig. 34) ; cut it up into a
large number of small pieces, and place these pieces by
means of an insulated rod at a distance of one centimetre from an electrical body which acts with unit of
force on an equal and like-constituted body at the
same distance.   Take the sum of the forces which
this last body exerts on the single pieces of the conductor.   The sum of these forces will be the quantity
of electricity on the whole conductor.   It remains the
same, whether we change the form and the size of the
conductor, or whether we bring it near or move it
away from a second electrical conductor, so long as we
keep it insulated, that is, do not discharge it.

A basis of reality for the notion of electric quantity seems also to present itself from another quarter.   If a current, that is, in the usual view, a definite
quantity of electricity per second, is sent through a
column of acidulated water ; in the direction of the
positive stream, hydrogen, but in the opposite direction, oxygen is liberated at the extremities of the column.   For a given quantity of electricity a given quantity of oxygen appears.   You may picture the column
of water as a column of hydrogen and a column of
oxygen, fitted into each other, and may say the electric
current is a chemical current and *vice versa*.   Although
this notion is more difficult to adhere to in the field of
statical electricity and with non-decomposable con-

ductors, its further development is by no means hopeless.

The concept quantity of electricity, thus, is not so aerial as might appear, but is able to conduct us with certainty through a multitude of varied phenomena, and is suggested to us by the facts in almost palpable form. We can collect electrical force in a body, measure it out with one body into another, carry it over from one body into another, just as we can collect a liquid in a vessel, measure it out with one vessel into another, or pour it from one into another.

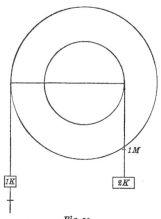

Fig. 35.

For the analysis of mechanical phenomena, a metrical notion, derived from experience, and bearing the designation *work*, has proved itself useful. A machine can be set in motion only when the forces acting on it can perform work.

Let us consider, for example, a wheel.and axle (Fig. 35) having the radii 1 and 2 metres, loaded respectively with the weights 2 and 1 kilogrammes. On turning the wheel and axle, the 1 kilogramme-weight, let us say, sinks two metres, while the 2 kilogramme-weight rises one metre. On both sides the product

$$\text{KGR. M. KGR. M.}$$
$$1 \times 2 = 2 \times 1.$$

is equal.   So long as this is so, the wheel and axle will
not move of itself.   But if we take such loads, or so
change the radii of the wheels, that this product (kgr.
$\times$ metre) on displacement is in excess on one side,
that side will sink.   As we see, this product is characteristic for mechanical events, and for this reason has
been invested with a special name, *work*.

In all mechanical processes, and as all physical
processes present a mechanical side, in all physical
processes, work plays a determinative part.   Electrical
forces, also, produce only changes in which work is performed.   To the extent that forces come into play in
electrical phenomena, electrical phenomena, be they
what they may, extend into the domain of mechanics
and are subject to the laws which hold in this domain.   The universally adopted measure of work,
now, is the product of the force into the distance
through which it acts, and in the C. G. S. system, the
unit of work is the action through one centimetre of
a force which would impart in one second to a
gramme-mass a velocity-increment of one centimetre,
that is, in round numbers, the action through a centimetre of a pressure equal to the weight of a milligramme.   From a positively charged body, electricity,
yielding to the force of repulsion and performing work,
flows off to the earth, providing conducting connexions
exist,   To a negatively charged body, on the other

hand, the earth under the same circumstances gives off positive electricity. The electrical work possible in the interaction of a body with the earth, characterises the electrical condition of that body. We will call the work which must be expended on the unit quantity of positive electricity to raise it from the earth to the body K the *potential* of the body K.*

We ascribe to the body K in the C. G. S. system the potential $+ 1$, if we must expend the unit of work to raise the positive electrostatic unit of electric quantity from the earth to that body; the potential -1, if we gain in this procedure the unit of work; the potential 0, if no work at all is performed in the operation.

The different parts of one and the same electrical conductor in electrical equilibrium have the same potential, for otherwise the electricity would perform work and move about upon the conductor, and equilibrium would not have existed. Different conductors of equal potential, put in connexion with one another, do not exchange electricity any more than bodies of equal temperature in contact exchange heat, or in connected vessels, in which the same pressures exist, liquids

* As this definition in its simple form is apt to give rise to misunderstandings, elucidations are usually added to it. It is clear that we cannot lift a quantity of electricity to K, without changing the distribution on K and the potential on K. Hence, the charges on K must be conceived as fixed, and so small a quantity raised that no appreciable change is produced by it. Taking the work thus expended as many times as the small quantity in question is contained in the unit of quantity, we shall obtain the potential. The potential of a body K may be briefly and precisely defined as follows : If we expend the element of work dW to raise the element of positive quantity dQ from the earth to the conductor, the potential of a conductor K will be given by $V = dW/dQ$.

flow from one vessel to the other. Exchange of electricity takes place only between conductors of different potentials, but in conductors of given form and position a definite difference of potential is necessary for a spark, which pierces the insulating air, to pass between them.

On being connected, every two conductors assume at once the same potential. With this the means is given of determining the potential of a conductor through the agency of a second conductor expressly adapted to the purpose called an electrometer, just as we determine the temperature of a body with a thermometer. The values of the potentials of bodies obtained in this way simplify vastly our analysis of their electrical behavior, as will be evident from what has been said.

Think of a positively charged conductor. Double all the electrical forces exerted by this conductor on a point charged with unit quantity, that is, double the quantity at each point, or what is the same thing, double the total charge. Plainly, equilibrium still subsists. But carry, now, the positive electrostatic unit towards the conductor. Everywhere we shall have to overcome double the force of repulsion we did before, everywhere we shall have to expend double the work. By doubling the charge of the conductor a double potential has been produced. Charge and potential go hand in hand, are proportional. Consequently, calling the total quantity of electricity of a conductor Q

and its potential V, we can write : $Q = CV$, where C stands for a constant, the import of which will be understood simply from noting that $C = Q/V$.* But the division of a number representing the units of quantity of a conductor by the number representing its units of potential tells us the quantity which falls to the share of the unit of potential. Now the number C here we call the capacity of a conductor, and have substituted, thus, in the place of the old relative determination of capacity, an absolute determination.†

In simple cases the connexion between charge, potential, and capacity is easily ascertained. Our conductor, let us say, is a sphere of radius r, suspended free in a large body of air. There being no other conductors in the vicinity, the charge q will then distribute itself uniformly upon the surface of the sphere, and simple geometrical considerations yield for its potential the expression $V = q/r$. Hence, $q/V = r$; that is, the capacity of a sphere is measured by its radius, and

---

* In this article the solidus or slant stroke is used for the usual fractional sign of division. Where plus or minus signs occur in the numerator or denominator, brackets or a vinculum is used.—*Tr.*

† A sort of agreement exists between the notions of thermal and electrical capacity, but the difference between the two ideas also should be carefully borne in mind. The thermal capacity of a body depends solely upon that body itself. The electrical capacity of a body K is influenced by all bodies in its vicinity, inasmuch as the charge of these bodies is able to alter the potential of K. To give, therefore, an unequivocal significance to the notion of the capacity (C) of a body K, C is defined as the relation Q/V for the body K in a certain given position of all neighboring bodies, and during connexion of all neighboring conductors with the earth. In practice the situation is much simpler. The capacity, for example, of a jar, the inner coating of which is almost enveloped by its outer coating, communicating with the ground, is not sensibly affected by charged or uncharged adjacent conductors.

in the C. G. S. system in centimetres.* It is clear also, since a potential is a quantity divided by a length, that a quantity divided by a potential must be a length.

Imagine (Fig. 36) a jar composed of two concentric conductive spherical shells of the radii r and r_1, having only air between them. Connecting the outside sphere with the earth, and charging the inside sphere by means of a thin, insulated wire passing through the first, with the quantity Q, we shall have $V = (r_1 - r)/(r_1 r) Q$, and for the capacity in this case

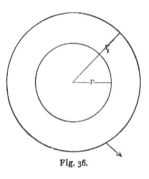

$(r_1 r)/(r_1 - r)$, or, to take a specific example, if $r = 16$ and $r_1 = 19$, a capacity of about 100 centimetres.

We shall now use these simple cases for illustrating the principle by which capacity and potential are determined. First, it is clear that we can use the

Fig. 36.

jar composed of concentric spheres with its known capacity as our unit jar and by means of this ascertain, in the manner above laid down, the capacity of any given jar F. We find, for example, that 37 discharges of this unit jar of the capacity 100, just charges the

* These formulæ easily follow from Newton's theorem that a homogeneous spherical shell, whose elements obey the law of the inverse squares, exerts no force whatever on points within it but acts on points without as if the whole mass were concentrated at its centre. The formulæ next adduced also flow from this proposition.

jar investigated at the same striking distance, that is, at the same potential. Hence, the capacity of the jar investigated is 3700 centimetres. The large battery of the Prague physical laboratory, which consists of sixteen such jars, all of nearly equal size, has a capacity, therefore, of something like 50,000 centimetres, or the capacity of a sphere, a kilometre in diameter, freely suspended in atmospheric space. This remark

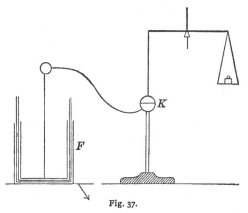

Fig. 37.

distinctly shows us the great superiority which Leyden jars possess for the storage of electricity as compared with common conductors. In fact, as Faraday pointed out, jars differ from simple conductors mainly by their great capacity.

For determining potential, imagine the inner coating of a jar *F*, the outer coating of which communicates with the ground, connected by a long, thin wire with a conductive sphere *K* placed free in a large atmospheric space, compared with whose dimensions the radius of the sphere vanishes. (Fig. 37.) The jar and the sphere assume at once the same potential. But on the surface of the sphere, if that be sufficiently far removed from all other conductors, a uniform layer of electricity will be found. If the sphere, having the radius r, contains the charge q, its potential is $V=q/r$. If the upper half of the sphere be severed from the lower half and equilibrated on a balance with one of whose beams it is connected by silk threads, the upper half will be repelled from the lower half with the force $P=q^2/8r^2=\frac{1}{8}V^2$. This repulsion P may be counterbalanced by additional weights placed on the beamend, and so ascertained. The potential is then $V=\sqrt{8P}$. *

That the potential is proportional to the square root of the force is not difficult to see. A doubling or trebling of the potential means that the charge of all the parts is doubled or trebled; hence their combined power of repulsion quadrupled or nonupled.

Let us consider a special case. I wish to produce the potential 40 on the sphere. What additional weight must I give to the half sphere in grammes that the force of repulsion shall maintain the balance in exact equilibrium? As a gramme weight is approximately

---

*The energy of a sphere of radius r charged with the quantity q is $\frac{1}{2}(q^2/r)$. If the radius increase by the space dr a loss of energy occurs, and the work done is $\frac{1}{2}(q^2/r^2)dr$. Letting p denote the uniform electrical pressure on unit of surface of the sphere, the work done is also $4r^2\pi p\,dr$. Hence $p=(1/8r^2\pi)(q^2/r^2)$. Subjected to the same superficial pressure on all sides, say in a fluid, our half sphere would be an equilibrium. Hence we must make the pressure p act on the surface of the great circle to obtain the effect on the balance, which is $r^2\pi p=\frac{1}{8}(q^2/r^2)=\frac{1}{8}V^2$.

equivalent to 1000 units of force, we have only the following simple example to work out : $40 \times 40 = 8 \times 1000 . x$, where x stands for the number of grammes. In round numbers we get $x = 0.2$ gramme. I charge the jar. The balance is deflected; I have reached, or rather passed, the potential 40, and you see when I discharge the jar the associated spark.*

The striking distance between the knobs of a machine increases with the difference of the potential, although not proportionately to that difference. The striking distance increases faster than the potential difference. For a distance between the knobs of one centimetre on this machine the difference of potential is 110. It can easily be increased tenfold. Of the tremendous differences of potential which occur in nature some idea may be obtained from the fact that the striking distances of lightning in thunder-storms is counted by miles. The differences of potential in galvanic batteries are considerably smaller than those of our machine, for it takes fully one hundred elements to give a spark of microscopic striking distance.

* * *

We shall now employ the ideas reached to shed some light upon another important relation between

---

* The arrangement described is for several reasons not fitted for the actual measurement of potential. Thomson's absolute electrometer is based upon an ingenious modification of the electrical balance of Harris and Volta. Of two large plane parallel plates, one communicates with the earth, while the other is brought to the potential to be measured. A small movable superficial portion f of this last hangs from the balance for the determination of the attraction P. The distance of the plates from each other being D we get $V = D\sqrt{8\pi P / f}$.

electrical and mechanical phenomena. We shall investigate what is the potential *energy*, or the *store of work*, contained in a charged conductor, for example, in a jar.

If we bring a quantity of electricity up to a conductor, or, to speak less pictorially, if we generate by work electrical force in a conductor, this force is able to produce anew the work by which it was generated. How great, now, is the energy or capacity for work of a conductor of known charge Q and known potential V?

Imagine the given charge Q divided into very small parts q, q_1, q_2 . . . ., and these little parts successively carried up to the conductor. The first very small quantity q is brought up without any appreciable work and produces by its presence a small potential $V_{,}$. To bring up the second quantity, accordingly, we must do the work $q, V_{,}$, and similarly for the quantities which follow the work $q_{,,} V_{,,}$, $q_{,,,} V_{,,,}$, and so forth. Now, as the potential rises proportionately to the quantities added until the value V is reached, we have, agreeably to the graphical representation of Fig. 38, for the total work performed,

$$W = \tfrac{1}{2} Q V,$$

which corresponds to the total energy of the charged conductor. Using the equation $Q = CV$, where C stands for capacity, we also have,

$$W = \tfrac{1}{2} C V^2, \text{ or } W = Q^2 / 2 C.$$

It will be helpful, perhaps, to elucidate this idea by an analogy from the province of mechanics. If we pump a quantity of liquid, Q, gradually into a cylindrical vessel (Fig. 39), the level of the liquid in the vessel will gradually rise. The more we have pumped in, the greater the pressure we must overcome, or the higher the level to which we must lift the liquid. The stored-up work is rendered again available when the heavy liquid Q, which reaches up to the level h, flows out. This work W corresponds to the fall of the whole

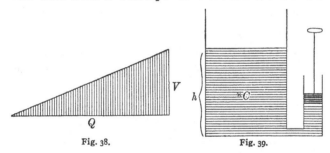

Fig. 38.   Fig. 39.

liquid weight Q, through the distance $h/2$ or through the altitude of its centre of gravity. We have

$$W = \tfrac{1}{2}Qh.$$

Further, since $Q = Kh$, or since the weight of the liquid and the height h are proportional, we get also

$$W = \tfrac{1}{2}Kh^2 \text{ and } W = Q^2/2K.$$

As a special case let us consider our jar. Its capacity is $C = 3700$, its potential $V = 110$; accordingly, its quantity $Q = CV = 407,000$ electrostatic units and its energy $W = \tfrac{1}{2}QV = 22,385,000$ C. G. S. units of work.

The unit of work of the C. G. S. system is not readily appreciable by the senses, nor does it well admit of representation, as we are accustomed to work with weights.   Let us adopt, therefore, as our unit of work the gramme-centimetre, or the gravitational pressure of a gramme-weight through the distance of a centimetre, which in round numbers is 1000 times greater than the unit assumed above; in this case, our numerical result will be approximately 1000 times smaller. Again, if we pass, as more familiar in practice, to the kilogramme-metre as our unit of work, our unit, the distance being increased a hundred fold, and the weight a thousand fold, will be 100,000 times larger.   The numerical result expressing the work done is in this case 100,000 times less, being in round numbers 0.22 kilogramme-metre.   We can obtain a clear idea of the work done here by letting a kilogramme-weight fall 22 centimetres.

This amount of work, accordingly, is performed on the charging of the jar, and on its discharge appears again, according to the circumstances, partly as sound, partly as a mechanical disruption of insulators, partly as light and heat, and so forth.

The large battery of the Prague physical laboratory, with its sixteen jars charged to equal potentials, furnishes, although the effect of the discharge is imposing, a total amount of work of only three kilogramme-metres.

In the development of the ideas above laid down we are not restricted to the method there pursued ; in fact, that method was selected only as one especially fitted to familiarise us with the phenomena. On the contrary, the connexion of the physical processes is so multifarious that we can come at the same event from very different directions. Particularly are electrical phenomena connected with all other physical events ; and so intimate is this connexion that we might justly call the study of electricity the theory of the general connexion of physical processes.

With respect to the principle of the conservation of energy which unites electrical with mechanical phe- nomena, I should like to point out briefly two ways of following up the study of this connexion.

A few years ago Professor Rosetti, taking an in- fluence-machine, which he set in motion by means of weights alternately in the electrical and non-electrical condition with the same velocities, determined the mechanical work expended in the two cases and was thus enabled, after deducting the work of friction, to ascertain the mechanical work consumed in the devel- opment of the electricity.

I myself have made this experiment in a modified, and, as I think, more advantageous form. Instead of determining the work of friction by special trial, I arranged my apparatus so that it was eliminated of it- self in the measurement and could consequently be neglected. The so-called fixed disk of the machine, the

axis of which is placed vertically, is suspended somewhat like a chandelier by three vertical threads of
equal lengths l at a distance r from the axis. Only
when the machine is excited does this fixed disk, which
represents a Prony's brake, receive, through its reciprocal action with the rotating disk, a deflexion α and a
moment of torsion which is expressed by $D = (Pr^2/l)\alpha$,
where P is the weight of the disk.* The angle α is
determined by a mirror set in the disk. The work expended in n rotations is given by $2\,n\,\pi\,D$.

If we close the machine, as Rosetti did, we obtain
a continuous current which has all the properties of a
very weak galvanic current; for example, it produces a
deflexion in a multiplier which we interpose, and so
forth. We can directly ascertain, now, the mechanical
work expended in the maintenance of this current.

If we charge a jar by means of a machine, the energy of the jar employed in the production of sparks,
in the disruption of the insulators, etc., corresponds
to a part only of the mechanical work expended, a
second part of it being consumed in the arc which
forms the circuit.† This machine, with the interposed
jar, affords in miniature a picture of the transference

---

*This moment of torsion needs a supplementary correction, on account of
the vertical electric attraction of the excited disks. This is done by changing
the weight of the disk by means of additional weights and by making a second
reading of the angles of deflexion.

†The jar in our experiment acts like an accumulator, being charged by a
dynamo machine. The relation which obtains between the expended and the
available work may be gathered from the following simple exposition. A
Holtz machine H (Fig. 40) is charging a unit jar L, which after n discharges
of quantity q and potential v, charges the jar F with the quantity Q at the po-

of force, or more properly of work. And in fact nearly the same laws hold here for the economical coefficient as obtain for large dynamo-machines.

Another means of investigating electrical energy is by its transformation into heat. A long time ago (1838), before the mechanical theory of heat had attained its present popularity, Riess performed experiments in this field with the help of his electrical air-thermometer or thermo-electrometer.

If the discharge be conducted through a fine wire passing through the globe of the air-thermometer, a development of heat is observed proportional to the expression above-discussed $W = \frac{1}{2} Q V$. Although the total energy has not yet been transformed into measurable heat by this means, inasmuch as a portion is left behind in the spark in the air outside the thermometer, still everything tends to show that the total

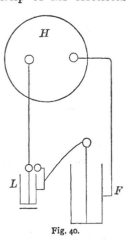

Fig. 40.

tential V. The energy of the unit-jar discharges is lost and that of the jar F alone is left. Hence the ratio of the available work to the total work expended is

$$\frac{\frac{1}{2} Q V}{\frac{1}{2} Q V + (n/2) q v} \text{ and as } Q = n q, \text{ also } \frac{V}{V + v}.$$

If, now, we interpose no unit jar, still the parts of the machine and the wires of conduction are themselves virtually such unit jars and the formula still subsists $V / \overline{V + \Sigma v}$, in which Σv represents the sum of all the successively introduced differences of potential in the circuit of connexion.

heat developed in all parts of the conductor and along all the paths of discharge is the equivalent of the work $\frac{1}{2}QV$.

It is not important here whether the electrical energy is transformed all at once or partly, by degrees. For example, if of two equal jars one is charged with the quantity Q at the potential V the energy present is $\frac{1}{2}QV$. If the first jar be discharged into the second, V, since the capacity is now doubled, falls to $V/2$. Accordingly, the energy $\frac{1}{4}QV$ remains, while $\frac{1}{4}QV$ is transformed in the spark of discharge into heat. The remainder, however, is equally distributed between the two jars so that each on discharge is still able to transform $\frac{1}{8}QV$ into heat.

*   *   *

We have here discussed electricity in the limited phenomenal form in which it was known to the inquirers before Volta, and which has been called, perhaps not very felicitously, "statical electricity." It is evident, however, that the nature of electricity is everywhere one and the same ; that a substantial difference between statical and galvanic electricity does not exist. Only the quantitative circumstances in the two provinces are so widely different that totally new aspects of phenomena may appear in the second, for example, magnetic effects, which in the first remained unnoticed, whilst, *vice versa*, in the second field statical attractions and repulsions are scarcely appreciable. As a fact, we can easily show the magnetic effect of the current of discharge of an influence-machine on the galvano-scope although we could hardly have made the original discovery of the magnetic effects with this current. The statical distant action of the wire poles of a galvanic element also would hardly have been noticed had not the phenomenon been known from a different quarter in a striking form.

If we wished to characterise the two fields in their chief and most general features, we should say that in the first, high potentials and small quantities come into play, in the second small potentials and large quantities. A jar which is discharging and a galvanic element deport themselves somewhat like an air-gun and the bellows of an organ. The first gives forth suddenly under a very high pressure a small quantity of air ; the latter liberates gradually under a very slight pressure a large quantity of air.

In point of principle, too, nothing prevents our retaining the electrostatical units in the domain of galvanic electricity and in measuring, for example, the strength of a current by the number of electrostatic units which flow per second through its cross-section. But this would be in a double aspect impractical. In the first place, we should totally neglect the magnetic facilities for measurement so conveniently offered by the current, and substitute for this easy means a method which can be applied only with difficulty and is not capable of great exactness. In the second place our units would be much too small, and we should find ourselves in the predicament of the astronomer who attempted to measure celestial distances in metres instead of in radii of the earth and the earth's orbit ; for the current which by the magnetic C. G. S. standard represents the unit, would require a flow of some 30,000,000,000 electrostatic units per second through its cross-section. Accordingly, different units must be adopted here. The development of this point, however, lies beyond my present task.

ON THE PRINCIPLE OF THE CON-
SERVATION OF ENERGY.*

IN a popular lecture, distinguished for its charming
simplicity and clearness, which Joule delivered in
the year 1847,† that famous physicist declares that the
living force which a heavy body has acquired by its
descent through a certain height and which it carries
with it in the form of the velocity with which it is impressed, is the *equivalent* of the attraction of gravity
through the space fallen through, and that it would be
"absurd" to assume that this living force could be destroyed without some restitution of that equivalent.
He then adds: "You will therefore be surprised to
hear that until very *recently* the universal opinion has
been that living force could be absolutely and irre-
vocably destroyed at any one's option." Let us add
that to-day, after forty-seven years, the *law of the conservation of energy*, wherever civilisation exists, is ac-

*Published in Vol. I, No. 5, of *The Monist*, October, 1894, being in part
a re-elaboration of the treatise *Ueber die Erhaltung der Arbeit*, Prague, 1872.

† *On Matter, Living Force, and Heat*, Joule: *Scientific Papers*, London,
1884, I, p. 265.

cepted as a fully established truth and receives the widest applications in all domains of natural science.

The fate of all momentous discoveries is similar. On their first appearance they are regarded by the majority of men as errors. J. R. Mayer's work on the principle of energy (1842) was rejected by the first physical journal of Germany; Helmholtz's treatise (1847) met with no better success; and even Joule, to judge from an intimation of Playfair, seems to have encountered difficulties with his first publication (1843). Gradually, however, people are led to see that the new view was long prepared for and ready for enunciation, only that a few favored minds had perceived it much earlier than the rest, and in this way the opposition of the majority is overcome. With proofs of the fruit-fulness of the new view, with its success, confidence in it increases. The majority of the men who employ it cannot enter into a deep-going analysis of it; for them, its success is its proof. It can thus happen that a view which has led to the greatest discoveries, like Black's theory of caloric, in a subsequent period in a province where it does not apply may actually become an obstacle to progress by its blinding our eyes to facts which do not fit in with our favorite conceptions. If a theory is to be protected from this dubious rôle, the grounds and motives of its evolution and existence must be examined from time to time with the utmost care.

The most multifarious physical changes, thermal,

electrical, chemical, and so forth, can be brought about by mechanical work. When such alterations are reversed they yield anew the mechanical work in exactly the quantity which was required for the production of the part reversed. This is the *principle of the conservation of energy*; "energy" being the term which has gradually come into use for that "indestructible something" of which the measure is mechanical *work*.

How did we acquire this idea? What are the sources from which we have drawn it? This question is not only of interest in itself, but also for the important reason above touched upon. The opinions which are held concerning the foundations of the law of energy still diverge very widely from one another. Many trace the principle to the impossibility of a perpetual motion, which they regard either as sufficiently proved by experience, or as self-evident. In the province of pure mechanics the impossibility of a perpetual motion, or the continuous production of *work* without some *permanent* alteration, is easily demonstrated. Accordingly, if we start from the theory that all physical processes are purely *mechanical* processes, motions of molecules and atoms, we embrace also, by this *mechanical* conception of physics, the impossibility of a perpetual motion in the *whole* physical domain. At present this view probably counts the most adherents. Other inquirers, however, are for accepting only a purely *experimental* establishment of the law of energy.

It will appear, from the discussion to follow, that *all* the factors mentioned have co operated in the development of the view in question; but that in addition to them a logical and purely formal factor, hitherto little considered, has also played a very important part.

I. THE PRINCIPLE OF THE EXCLUDED PERPETUAL MOTION.

The law of energy in its modern form is not identical with the principle of the excluded perpetual motion, but it is very closely related to it. The latter principle, however, is by no means new, for in the province of mechanics it has controlled for centuries the thoughts and investigations of the greatest thinkers. Let us convince ourselves of this by the study of a few historical examples.

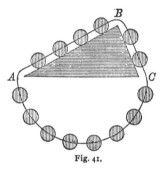

Fig. 41.

S. Stevinus, in his famous work *Hypomnemata mathematica*, Tom. IV, *De statica*, (Leyden, 1605, p. 34), treats of the equilibrium of bodies on inclined planes.

Over a triangular prism *A B C*, one side of which, *A C*, is horizontal, an endless cord or chain is slung, to which at equal distances apart fourteen balls of equal weight are attached, as represented in cross-section in Figure 41. Since we can imagine the lower symmetrical part of the cord *ABC* taken away, Stevinus concludes that the four balls on *A B* hold in equilibrium the two balls on *B C.* For if the equilibrium were for a moment disturbed, it could never subsist : the cord would keep moving round forever in the same direction,—we should have a perpetual motion. He says:

"But if this took place, our row or ring of balls would come once more into their original position, and from the same cause the eight globes to the left would again be heavier than the six to the right, and therefore those eight would sink a second time and these six rise, and all the globes would keep up, of themselves, *a continuous and unending motion, which is false.*" *

Stevinus, now, easily derives from this principle the laws of equilibrium on the inclined plane and numerous other fruitful consequences.

In the chapter "Hydrostatics" of the same work, page 114, Stevinus sets up the following principle : "Aquam datam, datum sibi intra aquam locum servare,"—a given mass of water preserves within water its given place.

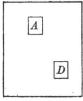

Fig. 42.

This principle is demonstrated as follows (see Fig. 42) :

"For, assuming it to be possible by natural means, let us suppose that *A* does not preserve the place assigned to it, but sinks down to *D.* This being posited, the water which succeeds *A* will,

* "Atqui hoc si sit, globorum series sive corona eundem situm cum priore habebit, eademque de causa octo globi sinistri ponderosiores erunt sex dextris, ideoque rursus octo illi descendent, sex illi ascendent, istique globi ex sese *continuum et aeternum motum efficient, quod est falsum.*"

for the same reason, also flow down to D ; A will be forced out of its place in D ; and thus this body of water, for the conditions in it are everywhere the same, *will set up a perpetual motion, which is absurd."* *

From this all the principles of hydrostatics are deduced.   On this occasion Stevinus also first develops the thought so fruitful for modern analytical mechanics that the equilibrium of a system is not destroyed by the addition of rigid connexions.   As we know, the principle of the conservation of the centre of gravity is now sometimes deduced from D'Alembert's principle with the help of that remark.   If we were to reproduce Stevinus's demonstration to-day, we should have to change it slightly.   We find no difficulty in imagining the cord on the prism possessed of unending uniform motion if all hindrances are thought away, but we should protest against the assumption of an accelerated motion or even against that of a uniform motion, if the resistances were not removed.   Moreover, for greater precision of proof, the string of balls might be replaced by a heavy homogeneous cord of infinite flexibility.   But all this does not affect in the least the historical value of Stevinus's thoughts.   It is a fact, Stevinus deduces apparently much simpler truths from the principle of an impossible perpetual motion.

* "A igitur, (si ullo modo per naturam fieri possit) locum sibi tributum non servato, ac delabatur in D ; quibus positis aqua quae ipsi A succedit eandem ob causam deffluet in D, eademque ab alia istinc expelletur, atque adeo aqua haec (cum ubique eadem ratio sit) *motum instituet perpetuum, quod absurdum fuerit."*

In the process of thought which conducted Galileo
to his discoveries at the end of the sixteenth century,
the following principle plays an important part, that
a body in virtue of the velocity acquired in its descent
can rise exactly as high as it fell. This principle,
which appears frequently and with much clearness in
Galileo's thought, is simply another form of the principle of excluded perpetual motion, as we shall see it
is also in Huygens.

Galileo, as we know, arrived at the law of uniformly
accelerated motion by *a priori* considerations, as that
law which was the "simplest and most natural," after
having first assumed a different law which he was compelled to reject. To verify his law he executed experiments with falling bodies on inclined planes, measuring the times of descent by the weights of the water
which flowed out of a small orifice in a large vessel.
In this experiment he assumes as a fundamental principle, that the velocity acquired in descent down an
inclined plane always corresponds to the vertical height
descended through, a conclusion which for him is the
immediate outcome of the fact that a body which has
fallen down one inclined plane can, with the velocity it
has acquired, rise on another plane of any inclination
only to the same vertical height. This principle of
the height of ascent also led him, as it seems, to the
law of inertia. Let us hear his own masterful words
in the *Dialogo terzo* (*Opere*, Padova, 1744, Tom. III).
On page 96 we read:

"I take it for granted that the velocities acquired by a body in descent down planes of different inclinations are equal if the heights of those planes are equal."*

Then he makes Salviati say in the dialogue :†

"What you say seems very probable, but I wish to go further and by an experiment so to increase the probability of it that it shall

* "Accipio, gradus velocitatis ejusdem mobilis super diversas planorum inclinationes acquisitos tunc esse aequales, cum eorundum planorum elevationes aequales sint."

† "Voi molto probabilmente discorrete, ma oltre al veri simile voglio con una esperienza crescer tanto la probabilità, che poco gli manchi all'agguagliarsi ad una ben necessaria dimostrazione. Figuratevi questo foglio essere una parete eretta al orizzonte, e da un chiodo fitto in essa pendere una palla di piombo d'un'oncia, o due, sospesa dal sottil filo *A B* lungo due, o tre braccia perpendicolare all' orizzonte, e nella parete segnate una linea orrizontale *D C* segante a squadra il perpendicolo *AB*, il quale sia lontano dalla parete due dita in circa, trasferendo poi il filo *A B* colla palla in *A C*, lasciata essa palla in libertà, la quale primier amente vedrete scendere descrivendo l'arco *C B D*, e di tanto trapassare il termine *B*, che scorrendo per l'arco *B D* sormonterà fino quasi alla segnata parallela *C D*, restando di per vernirvi per piccolissimo intervallo, toltogli il precisamente arrivarvi dall' impedimento dell'aria, e del filo. Dal che possiamo veracemente concludere, che l'impeto acquistato nel punto *B* dalla palla nello scendere per l'arco *C B*, fu tanto, che bastò a risospingersi per un simile arco *B D* alla medesima altezza ; fatta. e più volte reiterata cotale esperienza, voglio, che fiechiamo nella parete rasente al perpendicolo *A B* un chiodo come in *E*, ovvero in *F*, che sporga in fuori cinque, o sei dita, e questo acciocchè il filo *A C* tornando come prima a riportar la palla *C* per l'arco *C B*, giunta che ella sia in *B*, inoppando il filo nel chiodo *E*, sia costretta a camminare per la circonferenza *B G* descritta in torno al centro *E*, dal che vedremo quello, che potrà far quel medesimo impeto, che dianzi concepizo nel medesimo termine *B*, sospinse l'istesso mobile per l'arco *E D* all'altezza dell'orizzonale *C D*. Ora, Signori, voi vedrete con gusto condursi la palla all'orizzontale nel punto *G*, e l'istesso accadere, l'intoppo si metesse più basso. come in *F*, dove la palla descriverebbe l'arco *B F*, terminando sempre la sua salita precisamente nella linea *C D*, e quando l'intoppe del chiodo fusse tanto basso, che l'avanzo del filo sotto di lui non arivasse all'altezza di *C D* (il che accaderebbe, quando fusse più vicino all punto *B*, che al segamento dell' *A B* coll'orizzontale *C D*), allora il filo cavalcherebbe il chiodo, e segli avolgerebbe intorno. Questa esperienza non lascia luogo di dubitare della verità del supposto : imperocchè essendo li due archi *C B, D B* equali e similmento posti, l'acquisto di momento fatto per la scesa nell'arco *C B*, è il medesimo, che il fatto per la scesa dell'arco *D B* ; ma il momento acquistato in *B* per l'arco *C B* è potente a risospingere in su il medesimo mobile per l'arco *B D* ; adunque anco il momento acquistato nella scesa *D B* è eguale a quello, che sospigne l'istesso mobile pel medesimo arco da *B* in *D*, sicche universal-

amount almost to absolute demonstration. Suppose this sheet of
paper to be a vertical wall, and from a nail driven in it a ball of lead
weighing two or three ounces to hang by a very fine thread *AB* four
or five feet long. (Fig. 43.) On the wall mark a horizontal line *DC*
perpendicular to the vertical *AB*, which latter ought to hang about
two inches from the wall. If now the thread *AB* with the ball
attached take the position *AC* and the ball be let go, you will see
the ball first descend through the arc *CB* and passing beyond

B rise through the arc
BD almost to the level
of the line *CD*, being
prevented from reach-
ing it exactly by the re-
sistance of the air and
of the thread. From
this we may truly con-
clude that its impetus at
the point *B*, acquired by
its descent through the
arc *CB*, is sufficient to

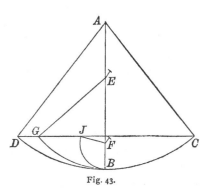

Fig. 43.

urge it through a similar arc *BD* to the same height. Having
performed this experiment and repeated it several times, let us
drive in the wall, in the projection of the vertical *AB*, as at *E* or
at *F*, a nail five or six inches long, so that the thread *AC*, carrying
as before the ball through the arc *CB*, at the moment it reaches
the position *AB*, shall strike the nail *E*, and the ball be thus compelled to move up the arc *BG* described about *E* as centre.
Then we shall see what the same impetus will here accomplish,
acquired now as before at the same point *B*, which then drove the

mente ogni momento acquistato per la scesa dun arco è eguale a quello, che
può far risalire l'istesso mobile pel medesimo arco : ma i momenti tutti che
fanno resalire per tutti gli archi *B D, B G, B J* sono eguali, poichè son fatti
dal istesso medesimo momento acquistato per la scesa *C B*, come mostra
l'esperienza : adunque tutti i momenti, che si acquistano per le scese negli
archi *D B, G B, J B* sono eguali."

same moving body through the arc *BD* to the height of the horizontal *CD*. Now gentlemen, you will be pleased to see the ball rise to the horizontal line at the point *G*, and the same thing also happen if the nail be placed lower as at *F*, in which case the ball would describe the arc *BJ*, always terminating its ascent precisely at the line *CD*. If the nail be placed so low that the length of thread below it does not reach to the height of *CD* (which would happen if *F* were nearer *B* than to the intersection of *AB* with the horizontal *CD*), then the thread will wind itself about the nail. This experiment leaves no room for doubt as to the truth of the supposition. For as the two arcs *CB*, *DB* are equal and similarly situated, the momentum acquired in the descent of the arc *CB* is the same as that acquired in the descent of the arc *DB*; but the momentum acquired at *B* by the descent through the arc *CB* is capable of driving up the same moving body through the arc *BD*; hence also the momentum acquired in the descent *DB* is equal to that which drives the same moving body through the same arc from *B* to *D*, so that in general every momentum acquired in the descent of an arc is equal to that which causes the same moving body to ascend through the same arc; but all the momenta which cause the ascent of all the arcs *BD*, *BG*, *BJ*, are equal since they are made by the same momentum acquired in the descent *CB*, as the experiment shows: therefore all the momenta acquired in the descent of the arcs *DB*, *GB*, *JB* are equal."

The remark relative to the pendulum may be applied to the inclined plane and leads to the law of inertia. We read on page 124 :*

---

* "Constat jam, quod mobile ex quiete in *A* descendens per *AB*, gradus acquirit velocitatis juxta temporis ipsius incrementum : gradum vero in *B* esse maximum acquisitorum, et suapte natura immutabiliter impressum, sublatis scilicet causis accelerationis novae, aut retardationis : accelerationis inquam, si adhuc super extenso plano ulterius progrederetur ; retardationis vero, dum super planum acclive *B C* fit reflexio : in horizontali autem *G H* aequabilis motus juxta gradum velocitatis ex *A* in *B* acquisitae in infinitum extenderetur.

"It is plain now that a movable body, starting from rest at *A* and descending down the inclined plane *A B*, acquires a velocity proportional to the increment of its time: the velocity possessed at *B* is the greatest of the velocities acquired, and by its nature immutably impressed, provided all causes of new acceleration or retardation are taken away: I say acceleration, having in view its possible further progress along the plane extended; retardation, in view of the possibility of its being reversed and made to mount the ascending plane *B C*. But in the horizontal plane *G H* its equable motion, according to its velocity as acquired in the descent from *A* to *B*, will be continued *ad infinitum*." (Fig. 44.)

Huygens, upon whose shoulders the mantel of Galileo fell, forms a sharper conception of the law of inertia

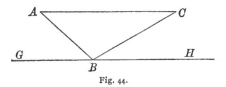

Fig. 44.

and generalises the principle respecting the heights of ascent which was so fruitful in Galileo's hands.   He employs the latter principle in the solution of the problem of the centre of oscillation and is perfectly clear in the statement that the principle respecting the heights of ascent is identical with the principle of the excluded perpetual motion.

The following important passages then occur (Hugenii, *Horologium oscillatorium, pars secunda*). *Hypotheses:*

"If gravity did not exist, nor the atmosphere obstruct the motions of bodies, a body would keep up forever the motion once impressed upon it, with equable velocity, in a straight line."*

In part four of the *Horologium, de centro oscillationis* we read :

"If any number of weights be set in motion by the force of gravity, the common centre of gravity of the weights as a whole cannot possibly rise higher than the place which it occupied when the motion began.

"That this hypothesis of ours may arouse no scruples, we will state that it simply imports, what no one has ever denied, that heavy bodies do not move *upwards.*—And truly if the devisers of the new machines who make such futile attempts to construct a perpetual motion would acquaint themselves with this principle, they could easily be brought to see their errors and to understand that the thing is utterly impossible by mechanical means." †

There is possibly a Jesuitical mental reservation contained in the words "mechanical means." One might be led to believe from them that Huygens held a non-mechanical perpetual motion for possible.

The generalisation of Galileo's principle is still more clearly put in Prop. IV of the same chapter :

"If a pendulum, composed of several weights, set in motion from rest, complete any part of its full oscillation, and from that

---

* "Si gravitas non esset, neque aër motui corporum officeret, unumquodque eorum, acceptum semel motum continuaturum velocitate aequabili, secundum lineam rectam."

† "Si pondera quotlibet, vi gravitatis suae, moveri incipiant ; non posse centrum gravitatis ex ipsis compositae altius, quam ubi incipiente motu reperiebatur, ascendere.

"Ipsa vero hypothesis nostra quominus scrupulum moveat, nihil aliud sibi velle ostendemus, quam, quod nemo unquam negavit, gravia nempe sursum non ferri.—Et sane, si hac eadem uti scirent novorum operum machinatores, qui motum perpetuum irrito conatu moliuntur, facile suos ipsi errores deprehenderent, intelligerentque rem eam mechanica ratione haud quaquam possibilem esse."

point onwards, the individual weights, with their common connexions dissolved, change their acquired velocities upwards and ascend as far as they can, the common centre of gravity of all will be carried up to the same altitude with that which it occupied before the beginning of the oscillation."*

On this last principle now, which is a generalisation, applied to a system of masses, of one of Galileo's ideas respecting a single mass and which from Huygens's explanation we recognise as the principle of excluded perpetual motion, Huygens grounds his theory of the centre of oscillation. Lagrange characterises this principle as precarious and is rejoiced at James Bernoulli's successful attempt, in 1681, to reduce the theory of the centre of oscillation to the laws of the lever, which appeared to him clearer. All the great inquirers of the seventeenth and eighteenth centuries broke a lance on this problem, and it led ultimately, in conjunction with the principle of virtual velocities, to the principle enunciated by D'Alembert in 1743 in his *Traité de dynamique*, though previously employed in a somewhat different form by Euler and Hermann.

Furthermore, the Huygenian principle respecting the heights of ascent became the foundation of the "law of the conservation of living force," as that was enunciated by John and Daniel Bernoulli and em-

---

* " Si pendulum e pluribus ponderibus compositum, atque e quiete dimissum, partem quamcunque oscillationis integrae confecerit, atque inde porro intelligantur pondera ejus singula, relicto communi vinculo, celeritates acquisitas sursum convertere, ac quousque possunt ascendere ; hoc facto centrum gravitatis ex omnibus compositae, ad eandem altitudinem reversum erit, quam ante inceptam oscillationem obtinebat."

ployed with such signal success by the latter in his *Hydrodynamics*. The theorems of the Bernoullis differ in form only from Lagrange's expression in the *Analytical Mechanics*.

The manner in which Torricelli reached his famous law of efflux for liquids leads again to our principle. Torricelli assumed that the liquid which flows out of the basal orifice of a vessel cannot by its velocity of efflux ascend to a greater height than its level in the vessel.

Let us next consider a point which belongs to pure mechanics, the history of the principle of *virtual motions* or *virtual velocities*. This principle was not first enunciated, as is usually stated, and as Lagrange also asserts, by Galileo, but earlier, by Stevinus. In his *Trochleostatica* of the above-cited work, page 72, he says :

"Observe that this axiom of statics holds good here :

"As the space of the body acting is to the space of the body acted upon, so is the power of the body acted upon to the power of the body acting."*

Galileo, as we know, recognised the truth of the principle in the consideration of the simple machines, and also deduced the laws of the equilibrium of liquids from it.

Torricelli carries the principle back to the properties of the centre of gravity. The condition control-

* " Notato autem hic illud staticum axioma etiam locum habere:
" Ut spatium agentis ad spatium patientis
Sic potentia patientis ad potentiam agentis."

ling equilibrium in a simple machine, in which power
and load are represented by weights, is that the common centre of gravity of the weights shall not sink.
Conversely, if the centre of gravity cannot sink equilibrium obtains, because heavy bodies of themselves
do not move upwards. In this form the principle of
virtual velocities is identical with Huygens's principle
of the impossibility of a perpetual motion.

John Bernoulli, in 1717, first perceived the universal
import of the principle of virtual movements for all
systems; a discovery stated in a letter to Varignon.
Finally, Lagrange gives a general demonstration of
the principle and founds upon it his whole *Analytical
Mechanics*. But this general demonstration is based
after all upon Huygens and Torricelli's remarks. Lagrange, as is known, conceives simple pulleys arranged
in the directions of the forces of the system, passes a
cord through these pulleys, and appends to its free
extremity a weight which is a common measure of all
the forces of the system. With no difficulty, now, the
number of elements of each pulley may be so chosen
that the forces in question shall be replaced by them.
It is then clear that if the weight at the extremity cannot sink, equilibrium subsists, because heavy bodies
cannot of themselves move upwards. If we do not go
so far, but wish to abide by Torricelli's idea, we may
conceive every individual force of the system replaced
by a special weight suspended from a cord passing
over a pulley in the direction of the force and attached

at its point of application.   Equilibrium subsists then
when the common centre of gravity of all the weights
together cannot sink.   The fundamental supposition
of this demonstration is plainly the impossibility of a
perpetual motion.

Lagrange tried in every way to supply a proof free
from extraneous elements and fully satisfactory, but
without complete success.   Nor were his successors
more fortunate.

The whole of mechanics, thus, is based upon an
idea, which, though unequivocal, is yet unwonted and
not coequal with the other principles and axioms of
mechanics.   Every student of mechanics, at some stage
of his progress, feels the uncomfortableness of this
state of affairs; every one wishes it removed; but sel-
dom is the difficulty stated in words.   Accordingly, the
zealous pupil of the science is highly rejoiced when he
reads in a master like Poinsot (*Théorie générale de
l'équilibre et du mouvement des systèmes*) the following
passage, in which that author is giving his opinion of
the *Analytical Mechanics* :

"In the meantime, because our attention in that work was first
wholly engrossed with the consideration of its beautiful develóp-
ment of mechanics, which seemed to spring complete from a single
formula, we naturally believed that the science was completed or
that it only remained to seek the demonstration of the principle of
virtual velocities.   But that quest brought back all the difficulties
that we had overcome by the principle itself.   That law so general,
wherein are mingled the vague and unfamiliar ideas of infinitely
small movements and of perturbations of equilibrium, only grew

obscure upon examination ; and the work of Lagrange supplying nothing clearer than the march of analysis, we saw plainly that the clouds had only appeared lifted from the course of mechanics because they had, so to speak, been gathered at the very origin of that science.

"At bottom, a general demonstration of the principle of virtual velocities would be equivalent to the establishment of the whole of mechanics upon a different basis : for the demonstration of a law which embraces a whole science is neither more nor less than the reduction of that science to another law just as general, but evident, or at least more simple than the first, and which, consequently, would render that useless." *

According to Poinsot, therefore, a proof of the principle of virtual movements is tantamount to a total rehabilitation of mechanics.

Another circumstance of discomfort to the mathematician is, that in the historical form in which mechanics at present exists, dynamics is founded on statics, whereas it is desirable that in a science which pretends to deductive completeness the more special

*"Cependant, comme dans cet ouvrage on ne fut d'abord attentif qu'à considérer ce beau développement de la mécanique qui semblait sortir tout entière d'une seule et même formule, on crut naturellement que la science etait faite, et qu'il ne restait plus qu'à chercher la démonstration du principe des vitesses virtuelles. Mais cette recherche ramena toutes les difficultés qu'on avait franchies par le principe même. Cette loi si générale, où se mêlent des idées vagues et étrangères de mouvements infinement petits et de perturbation d'équilibre, ne fit en quelque sorte que s'obsurcir à l'examen ; et le livre de Lagrange n'offrant plus alors rien de clair que la marche des calculs, on vit bien que les nuages n'avaient paru levé sur le cours de la mécanique que parcequ'ils étaient, pour ainsi dire, rassemblés à l'origine même de cette science.

"Une démonstration générale du principe des vitesses virtuelles devait au fond revenir a établir le mécanique entière sur une autre base : car la demonstration d'une loi qui embrasse toute une science ne peut être autre chose que la reduction de cette science à une autre loi aussi générale, mais évidente, ou du moins plus simple que la première, et qui partant la rende inutile."

statical theorems should be deducible from the more general dynamical principles.

In fact, a great master, Gauss, gave expression to this desire in his presentment of the principle of least constraint (Crelle's *Journal für reine und angewandte Mathematik*, Vol. IV, p. 233) in the following words: "Proper as it is that in the gradual development of a science, and in the instruction of individuals, the easy should precede the difficult, the simple the complex, the special the general, yet the mind, when once it has reached a higher point of view, demands the contrary course, in which all statics shall appear simply as a special case of mechanics." Gauss's own principle, now, possesses all the requisites of universality, but its difficulty is that it is not immediately intelligible and that Gauss deduced it with the help of D'Alembert's principle, a procedure which left matters where they were before.

Whence, now, is derived this strange part which the principle of virtual motion plays in mechanics? For the present I shall only make this reply. It would be difficult for me to tell the difference of impression which Lagrange's proof of the principle made on me when I first took it up as a student and when I subsequently resumed it after having made historical researches. It first appeared to me insipid, chiefly on account of the pulleys and the cords which did not fit in with the mathematical view, and whose action I would much rather have discovered from the principle

itself than have taken for granted. But now that I
have studied the history of the science I cannot imagine a more beautiful demonstration.

In fact, through all mechanics it is this self-same
principle of excluded perpetual motion which accomplishes almost all, which displeased Lagrange, but
which he still had to employ, at least tacitly, in his own
demonstration. If we give this principle its proper
place and setting, the paradox is explained.

The principle of excluded perpetual motion is thus
no new discovery; it has been the guiding idea, for
three hundred years, of all the great inquirers. But
the principle cannot properly be *based* upon mechanical perceptions. For long before the development of
mechanics the conviction of its truth existed and even
contributed to that development. Its power of conviction, therefore, must have more universal and
deeper roots. We shall revert to this point.

II. MECHANICAL PHYSICS.

It cannot be denied that an unmistakable tendency
has prevailed, from Democritus to the present day, to
explain *all* physical events *mechanically.* Not to mention earlier obscure expressions of that tendency we
read in Huygens the following : *

"There can be no doubt that light consists of the *motion* of a
certain substance. For if we examine its production, we find that

* *Traité de la lumière*, Leyden, 1690, p. 2.

here on earth it is principally fire and flame which engender it, both of which contain beyond doubt bodies which are in rapid movement, since they dissolve and destroy many other bodies more solid than they : while if we regard its effects, we see that when light is accumulated, say by concave mirrors, it has the property of combustion just as fire has, that is to say, it disunites the parts of bodies, which is assuredly a proof of *motion*, at least in the *true philosophy*, in which the causes of all natural effects are conceived as *mechanical* causes. Which in my judgment must be accomplished or all hope of ever understanding physics renounced." *

S. Carnot,† in introducing the principle of excluded perpetual motion into the theory of heat, makes the following apology :

"It will be objected here, perhaps, that a perpetual motion proved impossible for *purely mechanical actions*, is perhaps not so when the influence of *heat* or of electricity is employed.   But can phenomena of heat or electricity be thought of as due to anything else than to *certain motions of bodies*, and as such must they not be subject to the general laws of mechanics ?" ‡

---

* L'on ne sçaurait douter que la lumière ne consiste dans le *mouvement* de certaine matière.   Car soit qu'on regarde sa production, on trouve qu'içy sur la terre c'est principalement le feu et la flamme qui l'engendrent, lesquels contient sans doute des corps qui sont dans un mouvement rapide, puis qu'ils dissolvent et fon'dent plusieurs autres corps des plus solides : soit qu'on regarde ses effets, on voit que quand la lumière est ramasseé, comme par des miroires concaves, elle a la vertu de brûler comme le feu. c-est-à-dire qu'elle desunit les parties des corps ; ce qui marque assurément du *mouvement*, au moins dans la *vraye Philosophie*, dans laquelle on conçoit la cause de tous les effets naturels par des raisons de *mechanique*.   Ce qu'il faut faire à mon avis, ou bien renoncer à tout espérance de jamais rien comprendre dans la Physique."

† *Sur la puissance motrice du feu.*   (Paris, 1824.)

‡ "On objectra peut-être ici que le mouvement perpétuel, démontré impossible par les *seules actions mécaniques*, ne l'est peut-être pas lorsqu'on emploie l'influence soit de la *chaleur*, soit de l'électricité ; mais peut-on concevoir les phénomènes de la chaleur et de l'électricité comme dus à autre chose qu'à des *mouvements quelconques des corps* et comme tels ne doivent-ils pas être soumis aux lois générales de la mécanique ? "

These examples, which might be multiplied by quotations from recent literature indefinitely, show that a tendency to explain all things mechanically actually exists. This tendency is also intelligible. Mechanical events as simple motions in space and time best admit of observation and pursuit by the help of our highly organised senses. We reproduce mechanical processes almost without effort in our imagination. Pressure as a circumstance that produces motion is very familiar to us from daily experience. All changes which the individual personally produces in his environment, or humanity brings about by means of the arts in the world, are effected through the instrumentality of *motions*. Almost of necessity, therefore, motion appears to us as the most important physical factor. Moreover, mechanical properties may be discovered in all physical events. The sounding bell trembles, the heated body expands, the electrified body attracts other bodies. Why, therefore, should we not attempt to grasp all events under their mechanical aspect, since that is so easily apprehended and most accessible to observation and measurement? In fact, no objection *is* to be made to the attempt to elucidate the properties of physical events by mechanical *analogies*.

But modern physics has proceeded *very far* in this direction. The point of view which Wundt represents in his excellent treatise *On the Physical Axioms* is probably shared by the majority of physicists. The axioms of physics which Wundt sets up are as follows :

1. All natural causes are motional causes.

2. Every motional cause lies outside the object moved.

3. All motional causes act in the direction of the straight line of junction, and so forth.

4. The effect of every cause persists.

5. Every effect involves an equal countereffect.

6. Every effect is equivalent to its cause.

These principles might be studied properly enough as fundamental principles of mechanics. But when they are set up as axioms of physics, their enunciation is simply tantamount to a negation of all events except motion.

According to Wundt, all changes of nature are mere changes of place. All causes are motional causes (page 26). Any discussion of the philosophical grounds on which Wundt supports his theory would lead us deep into the speculations of the Eleatics and the Herbartians. Change of place, Wundt holds, is the *only* change of a thing in which a thing remains identical with itself. If a thing changed *qualitatively*, we should be obliged to imagine that something was annihilated and something else created in its place, which is not to be reconciled with our idea of the identity of the object observed and of the indestructibility of matter. But we have only to remember that the Eleatics encountered difficulties of exactly the same sort

in motion. Can we not also imagine that a thing is
destroyed in *one* place and in *another* an exactly similar thing created? After all, do we really know *more*
why a body leaves one place and appears in another,
than why a *cold* body grows *warm*? Granted that we
had a perfect knowledge of the mechanical processes
of nature, could we and should we, for that reason,
put out of the world all other processes that we do not
understand? On this principle it would really be the
simplest course to deny the existence of the whole
world. This is the point at which the Eleatics ultimately arrived, and the school of Herbart stopped
little short of the same goal.

Physics treated in this sense supplies us simply
with a diagram of the world, in which we do not know
reality again. It happens, in fact, to men who give
themselves up to this view for many years, that the
world of sense from which they start as a province of
the greatest familiarity, suddenly becomes, in their
eyes, the supreme "world-riddle."

Intelligible as it is, therefore, that the efforts of
thinkers have always been bent upon the "reduction
of all physical processes to the motions of atoms," it
must yet be affirmed that this is a chimerical ideal.
This ideal has often played an effective part in popular lectures, but in the workshop of the serious inquirer it has discharged scarcely the least function.
What has really been achieved in mechanical physics
is either the *elucidation* of physical processes by more

familiar *mechanical analogies*, (for example, the theories
of light and of electricity,) or the exact *quantitative*
ascertainment of the connexion of mechanical processes with other physical processes, for example, the
results of thermodynamics.

III. THE PRINCIPLE OF ENERGY IN PHYSICS.

We can know only from *experience* that mechanical
processes produce other physical transformations, or
vice versa. The attention was first directed to the connexion of mechanical processes, especially the performance of work, with changes of thermal conditions
by the invention of the steam-engine, and by its great
technical importance.   Technical interests and the
need of scientific lucidity meeting in the mind of S.
Carnot led to the remarkable development from which
thermodynamics flowed.   It is simply *an accident of
history* that the development in question was not connected with the practical applications of *electricity*.

In the determination of the maximum quantity of
work that, generally, a heat-machine, or, to take a
special case, a steam-engine, can perform with the
expenditure of a *given* amount of heat of combustion,
Carnot is guided by mechanical analogies. A body can
do work on being heated, by expanding under pressure.
But to do this the body must receive heat from a *hotter*
body.   Heat, therefore, to do work, must pass from a
hotter body to a colder body, just as water must fall
from a higher level to a lower level to put a mill-wheel

in motion. Differences of temperature, accordingly, represent forces able to do work exactly as do differences of height in heavy bodies. Carnot pictures to himself an ideal process in which no heat flows away unused, that is, without doing work. With a given expenditure of heat, accordingly, this process furnishes the maximum of work. An analogue of the process would be a mill-wheel which scooping its water out of a higher level would slowly carry it to a lower level without the loss of a drop. A peculiar property of the process is, that with the expenditure of the same work the water can be raised again exactly to its original level. This property of *reversibility* is also shared by the process of Carnot. His process also can be reversed by the expenditure of the same amount of work, and the heat again brought back to its original temperature level.

Suppose, now, we had *two* different reversible processes A, B, such that in A a quantity of heat, Q, flowing off from the temperature t_1 to the lower temperature t_2 should perform the work W, but in B under the same circumstances it should perform a greater quantity of work $W + W'$; then, we could join B in the sense assigned and A in the reverse sense into a *single* process. Here A would reverse the transformation of heat produced by B and would leave a surplus of work W', produced, so to speak, from nothing. The combination would present a perpetual motion.

With the feeling, now, that it makes little differ-

ence whether the mechanical laws are broken directly
or indirectly (by processes of heat), and convinced of
the existence of a *universal* law-ruled connexion of nature, Carnot here excludes for the first time from the
province of *general* physics the possibility of a perpetual motion. *But it follows, then, that the quantity
of work W, produced by the passage of a quantity of heat
Q from a temperature t_1 to a temperature t_2, is independent of the nature of the substances as also of the character of the process, so far as that is unaccompanied by
loss, but is wholly dependent upon the temperature t_1, t_2.*

This important principle has been fully confirmed
by the special researches of Carnot himself (1824), of
Clapeyron (1834), and of Sir William Thomson (1849),
now Lord Kelvin. The principle was reached *without
any assumption whatever* concerning the nature of heat,
simply by the exclusion of a perpetual motion. Carnot,
it is true, was an adherent of the theory of Black, according to which the sum-total of the quantity of heat
in the world is constant, but so far as his investigations have been hitherto considered the decision on
this point is-of no consequence. Carnot's principle
led to the most remarkable results. W. Thomson
(1848) founded upon it the ingenious idea of an "absolute" scale of temperature. James Thomson (1849)
conceived a Carnot process to take place with water
freezing under pressure and, therefore, performing
work. He discovered, thus, that the freezing point is
lowered 0·0075° Celsius by every additional atmos-

phere of pressure. This is mentioned merely as an example.

About twenty years after the publication of Carnot's book a further advance was made by J. R. Mayer and J. P. Joule. Mayer, while engaged as a physician in the service of the Dutch, observed, during a process of bleeding in Java, an unusual redness of the venous blood. In agreement with Liebig's theory of animal heat he connected this fact with the diminished loss of heat in warmer climates, and with the diminished expenditure of organic combustibles. The total expenditure of heat of a man at rest must be equal to the total heat of combustion. But since *all* organic actions, even the mechanical actions, must be set down to the credit of the heat of combustion, some connexion must exist between mechanical work and expenditure of heat.

Joule started from quite similar convictions concerning the galvanic battery. A heat of association equivalent to the consumption of the zinc can be made to appear in the galvanic cell. If a current is set up, a part of this heat appears in the conductor of the current. The interposition of an apparatus for the decomposition of water causes a part of this heat to disappear, which on the burning of the explosive gas formed, is reproduced. If the current runs an electromotor, a portion of the heat again disappears, which, on the consumption of the work by friction, again makes its appearance. Accordingly, both the heat

produced and the work produced, appeared to Joule
also as connected with the consumption of material.
The thought was therefore present, both to Mayer and
to Joule, of regarding heat and work as equivalent
quantities, so connected with each other that what is
lost in one form universally appears in another.   The
result of this was a *substantial* conception of heat and
of work, and *ultimately a substantial conception of energy.*   Here every physical change of condition is regarded as energy, the destruction of which generates
work or equivalent heat.   An electric charge, for example, is energy.

In 1842 Mayer had calculated from the physical
constants then universally accepted that by the disappearance of one kilogramme-calorie 365 kilogramme-
metres of work could be performed, and *vice versa.*
Joule, on the other hand, by a long series of delicate
and varied experiments beginning in 1843 ultimately
determined the mechanical equivalent of the kilogramme-calorie, more exactly, as 425 kilogramme-
metres.

If we estimate every change of physical condition
by the *mechanical work* which can be performed upon
the *disappearance* of that condition, and call this measure *energy*, then we can measure all physical changes
of condition, no matter how different they may be,
with the same common measure, and say: *the sum-
total of all energy remains constant.*   This is the form that
the principle of excluded perpetual motion received at

the hands of Mayer, Joule, Helmholtz, and W. Thomson in its extension to the whole domain of physics.

After it had been proved that heat must *disappear* if mechanical work was to be done at its expense, Carnot's principle could no longer be regarded as a complete expression of the facts. Its improved form was first given, in 1850, by Clausius, whom Thomson followed in 1851. It runs thus : " If a quantity of heat Q' is transformed into work in a reversible process, *another* quantity of heat Q of the absolute* temperature T_1 is lowered to the absolute temperature T_2." Here Q' is dependent only on Q, T_1, T_2, but is independent of the substances used and of the character of the process, so far as that is unaccompanied by loss. Owing to this last fact, it is sufficient to find the relation which obtains for some one well-known physical substance, say a gas, and some definite simple process. The relation found will be the one that holds generally. We get, thus,

$$\frac{Q'}{Q'+Q} = \frac{T_1 - T_2}{T_1} \quad . \quad . \quad . \quad . \quad . \quad . \quad . \quad (1)$$

that is, the quotient of the available heat Q' transformed into work divided by the sum of the transformed and transferred heats (the total sum used), the so-called *economical coefficient* of the process, is,

$$\frac{T_1 - T_2}{T_1}.$$

---

* By this is meant the temperature of a Celsius scale, the zero of which is 273° below the melting-point of ice.

IV. THE CONCEPTIONS OF HEAT.

When a cold body is put in contact with a warm body it is observed that the first body is warmed and that the second body is cooled. We may say that the first body is warmed *at the expense of* the second body. This suggests the notion of a thing, or heat-substance, which passes from the one body to the other. If two masses of water m, m', of unequal temperatures, be put together, it will be found, upon the rapid equalisation of the temperatures, that the respective changes of temperatures u and u' are inversely proportional to the masses and of opposite signs, so that the algebraical sum of the products is,

$$m u + m' u' = 0.$$

Black called the products $m u$, $m' u'$, which are decisive for our knowledge of the process, *quantities of heat.* We may form a very clear *picture* of these products by conceiving them with Black as measures of the quantities of some substance. But the essential thing is not this picture but the *constancy* of the sum of these products in simple processes of conduction. If a quantity of heat disappears at one point, an equally large quantity will make its appearance at some other point. The retention of this idea leads to the discovery of specific heat. Black, finally, perceives that also something else may appear for a vanished quantity of heat, namely : the fusion or vaporisation of a definite quantity of matter. He adheres here still to this favorite view, though with some freedom, and considers the vanished quantity of heat as still present, but as *latent*.

The generally accepted notion of a caloric, or heat-stuff, was strongly shaken by the work of Mayer and Joule. If the quantity of heat can be increased and diminished, people said, heat cannot be a substance, but must be a *motion*. The subordinate part of this statement has become much more popular than all the rest of the doctrine of energy. But we may convince ourselves that the motional conception of heat is now as unessential as was formerly its conception as a substance. Both ideas were favored or impeded solely by accidental historical circumstances. It does not follow that heat is not a substance from the fact that a mechanical equivalent exists for quantity of heat. We will make this clear by the following question which bright students have sometimes put to me. Is there a mechanical equivalent of electricity as there is a mechanical equivalent of heat? Yes, and no. There is no mechanical equivalent of *quantity* of electricity as there is an equivalent of *quantity* of heat, because the same quantity of electricity has a very different capacity for work, according to the circumstances in which it is placed; but there *is* a mechanical equivalent of electrical energy.

Let us ask another question. Is there a mechanical equivalent of water? No, there is no mechanical equivalent of quantity of water, but there is a mechanical equivalent of weight of water multiplied by its distance of descent.

When a Leyden jar is discharged and work thereby performed, we do not picture to ourselves that the quantity of electricity disappears as work is done, but we simply assume that the electricities come into different positions, equal quantities of positive and negative electricity being united with one another.

What, now, is the reason of this difference of view in our treatment of heat and of electricity? The reason is purely historical, wholly conventional, and, what is still more important, is wholly indifferent. I may be allowed to establish this assertion

In 1785 Coulomb constructed his torsion balance, by which he was enabled to measure the repulsion of electrified bodies. Suppose we have two small balls, A, B, which over their whole extent are similarly electrified. These two balls will exert on one another, at a certain distance r of their centres, a certain repulsion p. We bring into contact with B now a ball C, suffer both to be equally electrified, and then measure the repulsion of B from A and of C from A at the same distance r. The sum of these repulsions is again p. Accordingly something has remained constant. If we ascribe this effect to a substance, then we infer naturally its constancy. But the essential point of the exposition is the divisibility of the electric force p and not the simile of substance.

In 1838 Riess constructed his electrical air-thermometer (the thermoelectrometer). This gives a measure of the quantity of heat produced by the discharge of jars. This quantity of heat is not proportional to the quantity of electricity contained in the jar by Coulomb's measure, but if Q be this quantity and C be the capacity, is proportional to $Q^2/2C$, or, more simply still, to the energy of the charged jar. If, now, we discharge the jar completely through the thermometer, we obtain a certain quantity of heat, W. But if we make the discharge through the thermometer into a second jar, we obtain a quantity less than W. But we may obtain the remainder by completely discharging both jars through the air-thermometer, when it will again be proportional to the energy of the two jars. On the first, incomplete discharge, accordingly, a part of the electricity's capacity for work was lost.

When the charge of a jar produces heat its energy is changed and its value by Riess's thermometer is decreased. But by Coulomb's measure the quantity remains unaltered.

Now let us imagine that Riess's thermometer had been invented before Coulomb's torsion balance, which is not a difficult feat, since both inventions are independent of each other ; what would be more natural than that the "quantity" of electricity contained in a jar should be measured by the heat produced in the thermometer? But then, this so-called quantity of electricity would decrease on the production of heat or on the performance of work, whereas it now remains unchanged; in that case, therefore, electricity would not be a *substance* but a *motion*, whereas now it is still a substance. The reason, therefore, why we have other notions of electricity than we have of heat, is purely historical, accidental, and conventional.

This is also the case with other physical things. Water does not disappear when work is done. Why? Because we measure quantity of water with scales, just as we do electricity. But suppose the capacity of water for work were called quantity, and had to be measured, therefore, by a mill instead of by scales; then this quantity also would disappear as it performed the work. It may, now, be easily conceived that many substances are not so easily got at as water. In that case we should be unable to carry out the one kind of measurement with the scales whilst many other modes of measurement would still be left us.

In the case of heat, now, the historically established measure of "quantity" is accidentally the work-value of the heat. Accordingly, its quantity disappears when work is done. But that heat is not a substance follows from this as little as does the opposite conclusion that it is a substance. In Black's case the quantity of heat remains constant because the heat passes into no *other* form of energy.

If any one to-day should still wish to think of heat as a substance, we might allow that person this liberty with little ado. He would only have to assume that that which we call quantity of heat was the energy of a substance whose quantity remained unaltered, but whose energy changed. In point of fact we might much better say, in analogy with the other terms of physics, energy of heat, instead of quantity of heat.

When we wonder, therefore, at the discovery that heat is motion, we wonder at something that was never discovered. It is perfectly indifferent and possesses not the slightest scientific value, whether we think of heat as a substance or not. The fact is, heat behaves in some connexions like a substance, in others not. Heat is latent in steam as oxygen is latent in water.

V. THE CONFORMITY IN THE DEPORTMENT OF THE ENERGIES.

The foregoing reflexions will gain in lucidity from a consideration of the conformity which obtains in the behavior of all energies, a point to which I called attention long ago.*

A weight P at a height H_1 represents an energy $W_1 = P H_1$. If we suffer the weight to sink to a lower height H_2, during which work is done, and the work done is employed in the production of living force, heat, or an electric charge, in short, is transformed, then the energy $W_2 = P H_2$ is still *left*. The equation subsists

$$\frac{W_1}{H_1} = \frac{W_2}{H_2}, \quad \cdots \cdots \cdots \quad (2)$$

* I first drew attention to this fact in my treatise *Ueber die Erhaltung der Arbeit*, Prague, 1872. Before this, Zeuner had pointed out the analogy between mechanical and thermal energy. I have given a more extensive development of this idea in a communication to the *Sitzungsberichte der Wiener*

or, denoting the *transformed* energy by $W' = W_1 - W_2$ and the *transferred* energy, that transported to the lower level, by $W = W_2$,

$$\frac{W'}{W' + W} = \frac{H_1 - H_2}{H_1}, \quad \ldots \ldots \quad (3)$$

an equation in all respects analogous to equation (1) at page 165. The property in question, therefore, is by no means peculiar to heat. Equation (2) gives the relation between the energy taken from the higher level and that deposited on the lower level (the energy left behind); it says that these *energies* are proportional to the *heights of the levels.* An equation analogous to equation (2) may be set up for *every* form of energy; hence the equation which corresponds to equation (3), and so to equation (1), may be regarded as valid for every form. For electricity, for example, H_1, H_2 signify the potentials.

When we observe for the first time the agreement here indicated in the transformative law of the energies, it appears surprising and unexpected, for we do not perceive at once its reason. But to him who pursues the comparative historical method that reason will not long remain a secret.

Since Galileo, mechanical work, though long under a different name, has been a *fundamental concept* of mechanics, as also a very important notion in the applied sciences. The transformation of work into living force, and of living force into work, suggests directly the notion of energy—the idea having been first fruitfully employed by Huygens, although Thomas Young first called it by the *name* of "energy." Let us add to this the constancy of weight (really the constancy of mass) and we shall see that with respect to mechanical energy it is involved in the very definition of the term that the capacity for work or the potential energy of a weight is proportional to the height of the level at which it is, in the geometrical sense, and that it decreases on the lowering of the weight, on transformation, proportionally to the height of the level. The zero level here is wholly arbitrary. With this, equation (2) is given, from which all the other forms follow.

When we reflect on the tremendous start which mechanics had over the other branches of physics, it is not to be wondered at that the attempt was always made to apply the notions of that science wherever this was possible. Thus the notion of mass, for example, was imitated by Coulomb in the notion of quantity of electricity. In the further development of the theory of electricity, the notion of work was likewise immediately introduced in the theory of potential, and heights of electrical level were measured by the work of unit of quantity raised to that level. But with this the preceding equation with all its consequences is given for electrical energy. The case with the other energies was similar.

Thermal energy, however, appears as a special case. Only by the peculiar experiments mentioned could it be discovered that heat is an energy. But the measure of this energy by Black's quantity of heat is the outcome of fortuitous circumstances. In the first place, the accidental slight variability of the capacity for heat c with the temperature, and the accidental slight deviation of the usual thermometrical scales from the scale derived from *the tensions of gases*, brings it about that the notion "quantity of heat" can be set up and that the quantity of heat ct corresponding to a difference of temperature t is nearly proportional to the energy of the heat. It is a quite accidental historical circumstance that Amontons hit upon the idea of measuring temperature by the tension of a gas. It is certain in this that he did not think of the work of the heat.* But the numbers standing for temperature, thus, are made proportional to the tensions of gases, that is, to the work done by gases, with otherwise equal changes of volume. It thus happens that *temperature heights* and *level heights of work* are proportional to one another.

If properties of the thermal condition varying greatly from the tensions of gases had been chosen, this relation would have assumed very complicated forms, and the agreement between heat and the other energies above considered would not subsist. It is

* Sir William Thomson first consciously and intentionally introduced (1848, 1851) a *mechanical* measure of temperature similar to the electric measure of potential.

very instructive to reflect upon this point. A *natural law*, therefore, is not implied in the conformity of the behavior of the energies, but this conformity is rather conditioned by the uniformity of our modes of conception and is also partly a matter of good fortune.

VI. THE DIFFERENCES OF THE ENERGIES AND THE LIMITS OF THE PRINCIPLE OF ENERGY.

Of every quantity of heat Q which does work in a reversible process (one unaccompanied by loss) between the absolute temperatures T_1, T_2, only the portion

$$\frac{T_1 - T_2}{T_1}$$

is transformed into work, while the remainder is transferred to the lower temperature-level T_2. This transferred portion can, upon the reversal of the process, with the same expenditure of work, again be brought back to the level T_1. But if the process is not reversible, then more heat than in the foregoing case flows to the lower level, and the surplus can no longer be brought back to the higher level T_2 without some *special* expenditure. W. Thomson (1852), accordingly, drew attention to the fact, that in all non-reversible, that is, in all real thermal processes, quantities of heat are lost for mechanical work, and that accordingly a dissipation or waste of mechanical energy is taking place. In all cases, heat is only partially transformed into work, but frequently work is wholly transformed into heat. Hence, a tendency exists towards a diminution of the *mechanical* energy and towards an increase of the *thermal* energy of the world.

For a simple, closed cyclical process, accompanied by no loss, in which the quantity of heat Q_1 is taken from the level T_1, and the quantity Q_2 is deposited upon the level T_2, the following relation, agreeably to equation (2), exists,

$$-\frac{Q_1}{T_1} + \frac{Q_2}{T_2} = 0.$$

Similarly, for any number of compound reversible cycles Clausius finds the algebraical sum

$$\Sigma \frac{Q}{T} = 0,$$

and supposing the temperature to change continuously,

$$\int \frac{dQ}{T} = 0 \quad . \quad . \quad . \quad . \quad . \quad . \quad . \quad . \quad . \quad (4)$$

Here the elements of the quantities of heat deducted from a given level are reckoned negative, and the elements imparted to it, positive. If the process is not reversible, then expression (4), which Clausius calls *entropy*, increases. In actual practice this is always the case, and Clausius finds himself led to the statement:

1. That the energy of the world remains constant.

2. That the entropy of the world tends toward a maximum.

Once we have noted the above-indicated conformity in the behavior of different energies, the *peculiarity* of thermal energy here mentioned must strike us. Whence is this peculiarity derived, for, generally every energy passes only partly into another form, which is also true of thermal energy? The explanation will be found in the following.

Every transformation of a special kind of energy *A* is accompanied with a fall of potential of that particular kind of energy, including heat. But whilst for the other kinds of energy a transformation and therefore a loss of energy on the part of the kind sinking in potential is connected with the fall of the potential, with heat the case is different. Heat can suffer a fall of potential without sustaining a loss of energy, at least according to the customary mode of estimation. If a weight sinks, it must create perforce kinetic energy, or heat, or some other form of energy. Also, an electrical charge cannot suffer a fall of potential without loss of energy, i. e., without transformation. But heat can pass with a fall of temperature to a body of greater capacity and the same thermal energy still be preserved, so long as we regard *every quantity* of heat as energy. This it is that gives to heat, besides its property of energy, in many cases the character of a material *substance*, or quantity.

If we look at the matter in an unprejudiced light, we must ask if there is any scientific sense or purpose in still considering as energy a quantity of heat that can no longer be transformed into mechanical work, (for example, the heat of a closed equably warmed material system). The principle of energy certainly plays in this case a wholly superfluous rôle, which is assigned to it only from habit.* To maintain the principle of energy in the face of a knowledge of the dissipation or waste of mechanical energy, in the face of the increase of entropy is equivalent almost to the liberty which Black took when he regarded the heat of liquefaction as still present but latent.† It is to be remarked further, that the expressions "energy of the world" and "entropy of the world" are slightly permeated with scholasticism. Energy and entropy are *metrical* notions. What meaning can there be in applying these notions to a case in which they are not applicable, in which their values are not determinable?

If we could really determine the entropy of the world it would represent a true, absolute measure of time. In this way is best seen the utter tautology of a statement that the entropy of the world increases with the time. Time, and the fact that certain changes take place only in a definite sense, are one and the same thing.

---

*Compare my *Analysis of the Sensations*, Jena, 1886: English translation, Chicago, 1895.

† A better terminology appears highly desirable in the place of the usual misleading one. Sir William Thomson (1852) appears to have felt this need, and it has been clearly expressed by F. Wald (1889). We should call the work which corresponds to a vanished quantity of heat its mechanical substitution-value; while that work which can be *actually* performed in the passage of a thermal condition A to a condition B, alone deserves the name of the *energy-value* of this change of condition. In this way the *arbitrary* substantial conception of the processes would be preserved and misapprehensions forestalled.

VII. THE SOURCES OF THE PRINCIPLE OF ENERGY.

We are now prepared to answer the question, What are the sources of the principle of energy? All knowledge of nature is derived in the last instance from experience. In this sense they are right who look upon the principle of energy as a result of experience.

Experience teaches that the sense-elements $\alpha\beta\gamma\delta$.... into which the world may be decomposed, are subject to change. It tells us further, that certain of these elements are *connected* with other elements, so that they appear and disappear together; or, that the appearance of the elements of one class is connected with the disappearance of the elements of the other class. We will avoid here the notions of cause and effect because of their obscurity and equivocalness. The result of experience may be expressed as follows: *The sensuous elements of the world* ($\alpha\,\beta\,\gamma\,\delta$ . . . .) *show themselves to be interdependent*. This interdependence is best represented by some such conception as is in geometry that of the mutual dependence of the sides and angles of a triangle, only much more varied and complex.

As an example, we may take a mass of gas enclosed in a cylinder and possessed of a definite volume (α), which we change by a pressure (β) on the piston, at the same time feeling the cylinder with our hand and

receiving a sensation of heat (γ).   Increase of pressure diminishes the volume and increases the sensation of heat.

   The various facts of experience are not in all respects alike.   Their common sensuous elements are
placed in relief by a process of abstraction and thus
impressed upon the memory.   In this way the expression is obtained of the features of *agreement* of extensive
groups of facts.   The simplest sentence which we can
utter is, by the very nature of language, an abstraction
of this kind.   But account must also be taken of the
differences of related facts.   Facts may be so nearly related as to contain the same kind of $\alpha\beta\gamma$ . . ., but the
relation be such that the $\alpha\beta\gamma$ . . . of the one differ
from the $\alpha\beta\gamma$ . . . of the other only by the number of
equal parts into which they can be divided.   Such
being the case, if rules can be given for deducing *from
one another* the numbers which are the measures of
these $\alpha\beta\gamma$ . . ., then we possess in such rules the *most
general* expression of a group of facts, as also that expression which corresponds to all its differences.   This
is the goal of quantitative investigation.

   If this goal be reached what we have found is that
between the $\alpha\beta\gamma$ . . . of a group of facts, or better, between the numbers which are their measures, a num-
ber of equations exists.   The simple fact of change
brings it about that the number of these equations
must be smaller than the number of the $\alpha\beta\gamma$ . . .   If
the former be smaller by one than the latter, then one

portion of the $\alpha\beta\gamma\ldots$ is *uniquely* determined by the other portion.

The quest of relations of this last kind is the most important function of special experimental research, because we are enabled by it to complete in thought facts that are only partly given. It is self-evident that only experience can ascertain that between the $\alpha\beta\gamma\ldots$ relations exist and of what kind they are. Further, only experience can tell that the relations that exist between the $\alpha\beta\gamma\ldots$ are such that changes of them can be reversed. If this were not the fact all occasion for the enunciation of the principle of energy, as is easily seen, would be wanting. In experience, therefore, is buried the ultimate well spring of all knowledge of nature, and consequently, in this sense, also the ultimate source of the principle of energy.

But this does not exclude the fact that the principle of energy has also a logical root, as will now be shown. Let us assume on the basis of experience that one group of sensuous elements $\alpha\beta\gamma\ldots$ determines *uniquely* another group $\lambda\mu\nu\ldots$ Experience further teaches that changes of $\alpha\beta\gamma\ldots$ can be *reversed*. It is then a logical consequence of this observation, that every time that $\alpha\beta\gamma\ldots$ assume the same values this is also the case with $\lambda\mu\nu\ldots$ Or, that purely *periodical* changes of $\alpha\beta\gamma\ldots$ can produce no *permanent* changes of $\lambda\mu\nu\ldots$ If the group $\lambda\mu\nu\ldots$ is a mechanical group, then a perpetual motion is excluded.

It will be said that this is a vicious circle, which we will grant. But psychologically, the situation is essentially different, whether I think simply of the unique determination and reversibility of events, or whether I exclude a perpetual motion. The attention takes in the two cases different directions and diffuses light over different sides of the question, which logically of course are necessarily connected.

Surely that firm, logical setting of the thoughts noticeable in the great inquirers, Stevinus, Galileo, and the rest, which, consciously or instinctively, was supported by a fine feeling for the slightest contradictions, has no other purpose than to limit the bounds of thought and so exempt it from the possibility of error. In this, therefore, the logical root of the principle of excluded perpetual motion is given, namely, in that universal conviction which existed even before the development of mechanics and co-operated in that development.

It is perfectly natural that the principle of excluded perpetual motion should have been first developed in the simple domain of pure mechanics. Towards the transference of that principle into the domain of general physics the idea contributed much that all physical phenomena are mechanical phenomena. But the foregoing discussion shows how little essential this notion is. The issue really involved is the recognition of a general interconnexion of nature. This once established, we see with Carnot that it is indifferent whether the mechanical laws are broken directly or circuitously.

The principle of the excluded perpetual motion is very closely related to the modern principle of energy, but it is not identical with it, for the latter is to be deduced from the former only by means of a definite *formal conception.* As may be seen from the preceding exposition, the perpetual motion can be excluded without our employing or possessing the notion of *work.* The modern principle of energy results primarily from a *substantial* conception of work and of every change of physical condition which by being reversed produces work. The strong need of such a conception, which is by no means necessary, but in a formal sense is very convenient and lucid, is exhibited in the case of J. R. Mayer and Joule. It was before remarked that this conception was suggested to both inquirers by the observation that both the production of heat and the production of mechanical work were connected with an expenditure of substance. Mayer says: "Ex nihilo nil fit," and in another place, "The creation or destruction of a force (work) lies without the province of human activity." In Joule we find this passage: "It is manifestly *absurd* to suppose that the powers with which God has endowed matter can be destroyed."

Some writers have observed in such statements the attempt at a *metaphysical* establishment of the doctrine of energy. But we see in them simply the formal need of a simple, clear, and living grasp of the facts, which

receives its development in practical and technical life,
and which we carry over, as best we can, into the
province of science. As a fact, Mayer writes to Grie-
singer: "If, finally, you ask me how I became involved
in the whole affair, my answer is simply this: Engaged
during a sea voyage almost exclusively with the study
of physiology, I discovered the new theory for the
sufficient reason that I *vividly felt the need of it.*"

The substantial conception of work (energy) is by
no means a necessary one. And it is far from true that
the problem is solved with the recognition of the need
of such a conception. Rather let us see how Mayer
gradually endeavored to satisfy that need. He first
regards quantity of motion, or momemtum, $m\,v$, as the
equivalent of work, and did not light, until later, on
the notion of living force ($m\,v^2/2$). In the province
of electricity he was unable to assign the expression
which is the equivalent of work. This was done later
by Helmholtz. The formal need, therefore, is *first*
present, and our conception of nature is subsequently
gradually *adapted* to it.

The laying bare of the experimental, logical, and
formal root of the present principle of energy will perhaps contribute much to the removal of the mysticism
which still clings to this principle. With respect to
our formal need of a very simple, palpable, substantial conception of the processes in our environment, it
remains an open question how far nature corresponds
to that need, or how far we can satisfy it. In one

phase of the preceding discussions it would seem as if the substantial notion of the principle of energy, like Black's material conception of heat, has its natural limits in facts, beyond which it can only be artificially adhered to.

THE ECONOMICAL NATURE OF
PHYSICAL INQUIRY.*

WHEN the human mind, with its limited powers, attempts to mirror in itself the rich life of the world, of which it is itself only a small part, and which it can never hope to exhaust, it has every reason for proceeding economically. Hence that tendency, expressed in the philosophy of all times, to compass by a few organic thoughts the fundamental features of reality. "Life understands not death, nor death life." So spake an old Chinese philosopher. Yet in his unceasing desire to diminish the boundaries of the incomprehensible, man has always been engaged in attempts to understand death by life and life by death.

Among the ancient civilised peoples, nature was filled with demons and spirits having the feelings and desires of men. In all essential features, this animistic view of nature, as Tylor† has aptly termed it, is shared in common by the fetish-worshipper of modern Africa

---

* An address delivered before the anniversary meeting of the Imperial Academy of Sciences, at Vienna, May 25, 1882.

† *Primitive Culture.*

and the most advanced nations of antiquity. As a theory of the world it has never completely disappeared. The monotheism of the Christians never fully overcame it, no more than did that of the Jews. In the belief in witchcraft and in the superstitions of the sixteenth and seventeenth centuries, the centuries of the rise of natural science, it assumed frightful pathological dimensions. Whilst Stevinus, Kepler, and Galileo were slowly rearing the fabric of modern physical science, a cruel and relentless war was waged with firebrand and rack against the devils that glowered from every corner. To-day even, apart from all survivals of that period, apart from the traces of fetishism which still inhere in our physical concepts,* those very ideas still covertly lurk in the practices of modern spiritualism.

By the side of this animistic conception of the world, we meet from time to time, in different forms, from Democritus to the present day, another view, which likewise claims exclusive competency to comprehend the universe. This view may be characterised as the *physico-mechanical* view of the world. To-day, that view holds, indisputably, the first place in the thoughts of men, and determines the ideals and the character of our times. The coming of the mind of man into the full consciousness of its powers, in the eighteenth century, was a period of genuine disillusionment. It produced the splendid precedent of a life

* Tylor, *loc. cit.*

really worthy of man, competent to overcome the old barbarism in the practical fields of life ; it created the *Critique of Pure Reason,* which banished into the realm of shadows the sham-ideas of the old metaphysics ; it pressed into the hands of the mechanical philosophy the reins which it now holds.

The oft-quoted words of the great Laplace,* which I will now give, have the ring of a jubilant toast to the scientific achievements of the eighteenth century : "A mind to which were given for a single instant all the forces of nature and the mutual positions of all its masses, if it were otherwise powerful enough to sub-ject these problems to analysis, could grasp, with a single formula, the motions of the largest masses as well as of the smallest atoms ; nothing would be un-certain for it ; the future and the past would lie re-vealed before its eyes." In writing these words, La-place, as we know, had also in mind the atoms of the brain. That idea has been expressed more forcibly still by some of his followers, and it is not too much to say that Laplace's ideal is substantially that of the great majority of modern scientists.

Gladly do we accord to the creator of the *Méca-nique céleste* the sense of lofty pleasure awakened in him by the great success of the Enlightenment, to which we too owe our intellectual freedom. But to-day, with minds undisturbed and before *new* tasks, it

* *Essai philosophique sur les probabilités.* 6th Ed. Paris, 1840, p. 4. The necessary consideration of the initial velocities is lacking in this formulation.

becomes physical science to secure itself against self-deception by a careful study of its character, so that it can pursue with greater sureness its true objects. If I step, therefore, beyond the narrow precincts of my specialty in this discussion, to trespass on friendly neighboring domains, I may plead in my excuse that the subject-matter of knowledge is common to all domains of research, and that fixed, sharp lines of demarcation cannot be drawn.

The belief in occult magic powers of nature has gradually died away, but in its place a new belief has arisen, the belief in the magical power of science. Science throws her treasures, not like a capricious fairy into the laps of a favored few, but into the laps of all humanity, with a lavish extravagance that no legend ever dreamt of! Not without apparent justice, therefore, do her distant admirers impute to her the power of opening up unfathomable abysses of nature, to which the senses cannot penetrate. Yet she who came to bring light into the world, can well dispense with the darkness of mystery, and with pompous show, which she needs neither for the justification of her aims nor for the adornment of her plain achievements.

The homely beginnings of science will best reveal to us its simple, unchangeable character. Man acquires his first knowledge of nature half-consciously and automatically, from an instinctive habit of mimicking and forecasting facts in thought, of supplementing sluggish experience with the swift wings of thought, at first only for his material welfare. When he hears a noise in the underbrush he constructs there, just as the animal does, the enemy which he fears ; when he sees a certain rind he forms mentally the image of the fruit which he is in search of ; just as we mentally associate a certain kind of matter with a certain line in the spectrum or an electric spark with the friction of a piece of glass. A knowledge of causality in this form certainly reaches far below the level of Schopenhauer's pet dog, to whom it was ascribed. It probably exists in the whole animal world, and confirms that great thinker's statement regarding the will which created the intellect for its purposes. These primitive psychical functions are rooted in the economy of our organism not less firmly than are motion and digestion. Who would deny that we feel in them, too, the elemental power of a long practised logical and physiological activity, bequeathed to us as an heirloom from our forefathers?

Such primitive acts of knowledge constitute to-day the solidest foundation of scientific thought. Our instinctive knowledge, as we shall briefly call it, by virtue of the conviction that we have consciously and intentionally contributed nothing to its formation, confronts us with an authority and logical power which consciously acquired knowledge even from familiar sources and of easily tested fallibility can never possess. All so-called axioms are such instinctive knowledge. Not consciously gained knowledge alone, but powerful intellectual instinct, joined with vast conceptive powers, constitute the great inquirer. The greatest advances of science have always consisted in some successful formulation, in clear, abstract, and communicable terms, of what was instinctively known long before, and of thus making it the permanent property of humanity. By Newton's principle of the equality of pressure and counterpressure, whose truth all before him had felt, but which no predecessor had abstractly formulated, mechanics was placed by a single stroke on a higher level. Our statement might also be historically justified by examples from the scientific labors of Stevinus, S. Carnot, Faraday, J. R. Mayer, and others.

All this, however, is merely the soil from which science starts. The first real beginnings of science appear in society, particularly in the manual arts, where the necessity for the communication of experience arises. Here, where some new discovery is to be described and related, the compulsion is first felt of clearly defining in consciousness the important and essential features of that discovery, as many writers can testify. The aim of instruction is simply the saving of experience; the labor of one man is made to take the place of that of another.

The most wonderful economy of communication is found in language. Words are comparable to type, which spare the repetition of written signs and thus serve a multitude of purposes; or to the few sounds of which our numberless different words are composed.

Language, with its helpmate, conceptual thought, by fixing the essential and rejecting the unessential, constructs its rigid pictures of the fluid world on the plan of a mosaic, at a sacrifice of exactness and fidelity but with a saving of tools and labor.   Like a piano-player with previously prepared sounds, a speaker excites in his listener thoughts previously prepared, but fitting many cases, which respond to the speaker's summons with alacrity and little effort.

The principles which a prominent political economist, E. Hermann,* has formulated for the economy of the industrial arts, are also applicable to the ideas of common life and of science. The economy of language is augmented, of course, in the terminology of science. With respect to the economy of written intercourse there is scarcely a doubt that science itself will realise that grand old dream of the philosophers of a Universal Real Character.   That time is not far distant. Our numeral characters, the symbols of mathematical analysis, chemical symbols, and musical notes, which might easily be supplemented by a system of color-signs, together with some phonetic alphabets now in use, are all beginnings in this direction.   The logical extension of what we have, joined with a use of the ideas which the Chinese ideography furnishes us, will render the special invention and promulgation of a Universal Character wholly superfluous.

The communication of scientific knowledge always

* *Principien der Wirthschaftslehre*, Vienna, 1873.

involves description, that is, a mimetic reproduction of facts in thought, the object of which is to replace and save the trouble of new experience. Again, to save the labor of instruction and of acquisition, concise, abridged description is sought. This is really all that natural laws are. Knowing the value of the acceleration of gravity, and Galileo's laws of descent, we possess simple and compendious directions for reproducing in thought all possible motions of falling bodies. A formula of this kind is a complete substitute for a full table of motions of descent, because by means of the formula the data of such a table can be easily constructed at a moment's notice without the least burdening of the memory.

No human mind could comprehend all the individual cases of refraction. But knowing the index of refraction for the two media presented, and the familiar law of the sines, we can easily reproduce or fill out in thought every conceivable case of refraction. The advantage here consists in the disburdening of the memory; an end immensely furthered by the written preservation of the natural constants. More than this comprehensive and condensed report about facts is not contained in a natural law of this sort. In reality, the law always contains less than the fact itself, because it does not reproduce the fact as a whole but only in that aspect of it which is important for us, the rest being either intentionally or from necessity omitted. Natural laws may be likened to intellectual type of a higher order, partly movable, partly stereotyped, which last on new editions of experience may become downright impediments.

When we look over a province of facts for the first time, it appears to us diversified, irregular, confused, full of contradictions. We first succeed in grasping only single facts, unrelated with the others. The province, as we are wont to say, is not *clear*. By and by we discover the simple, permanent elements of the mosaic, out of which we can mentally construct the whole province. When we have reached a point where we can discover everywhere the same facts, we no longer feel lost in this province ; we comprehend it without effort ; it is *explained* for us.

Let me illustrate this by an example. As soon as we have grasped the fact of the rectilinear propagation of light, the regular course of our thoughts stumbles at the phenomena of refraction and diffraction. As soon as we have cleared matters up by our index of refraction we discover that a special index is necessary for each color. Soon after we have accustomed ourselves to the fact that light added to light increases its intensity, we suddenly come across a case of total darkness produced by this cause. Ultimately, however, we see everywhere in the overwhelming multifariousness of optical phenomena the fact of the spatial and temporal periodicity of light, with its velocity of propagation dependent on the medium and the period. This tendency of obtaining a survey of a given province with the least expenditure of thought, and of representing all its facts by some one single mental process, may be justly termed an economical one.

The greatest perfection of mental economy is attained in that science which has reached the highest formal development, and which is widely employed in physical inquiry, namely, in mathematics. Strange as it may sound, the power of mathematics rests upon its evasion of all unnecessary thought and on its wonderful saving of mental operations. Even those arrangement-signs which we call numbers are a system of marvellous simplicity and economy. When we employ the multiplication-table in multiplying numbers of several places, and so use the results of old operations of counting instead of performing the whole of each operation anew; when we consult our table of logarithms, replacing and saving thus new calculations by old ones already performed; when we employ determinants instead of always beginning afresh the solution of a system of equations; when we resolve new integral expressions into familiar old integrals; we see in this simply a feeble reflexion of the intellectual activity of a Lagrange or a Cauchy, who, with the keen discernment of a great military commander, substituted for new operations whole hosts of old ones. No one will dispute me when I say that the most elementary as well as the highest mathematics are economically-ordered experiences of counting, put in forms ready for use.

In algebra we perform, as far as possible, all numerical operations which are identical in form once for all, so that only a remnant of work is left for the individual case. The use of the signs of algebra and analysis, which are merely symbols of operations to be performed, is due to the observation that we can materially disburden the mind in this way and spare its powers for more important and more difficult duties, by imposing all mechanical operations upon the hand. One result of this method, which attests its economical character, is the construction of calculating machines. The mathematician Babbage, the inventor of the difference-engine, was probably the first who clearly perceived this fact, and he touched upon it, although only cursorily, in his work, *The Economy of Manufactures and Machinery.*

The student of mathematics often finds it hard to throw off the uncomfortable feeling that his science, in the person of his pencil, surpasses him in intelligence, —an impression which the great Euler confessed he often could not get rid of. This feeling finds a sort of justification when we reflect that the majority of the ideas we deal with were conceived by others, often centuries ago. In great measure it is really the intelligence of other people that confronts us in science. The moment we look at matters in this light, the uncanniness and magical character of our impressions cease, especially when we remember that we can think over again at will any one of those alien thoughts.

Physics is experience, arranged in economical order. By this order not only is a broad and comprehensive view of what we have rendered possible, but also
the defects and the needful alterations are made manifest, exactly as in a well-kept household. Physics
shares with mathematics the advantages of succinct
description and of brief, compendious definition, which
precludes confusion, even in ideas where, with no apparent burdening of the brain, hosts of others are contained. Of these ideas the rich contents can be produced at any moment and displayed in their full per-
ceptual light. Think of the swarm of well-ordered notions pent up in the idea of the potential. Is it wonderful that ideas containing so much finished labor should
be easy to work with?

Our first knowledge, thus, is a product of the
economy of self-preservation. By communication, the
experience of *many* persons, individually acquired at
first, is collected in *one*. The communication of
knowledge and the necessity which every one feels of
managing his stock of experience with the least expenditure of thought, compel us to put our knowledge in
economical forms. But here we have a clue which
strips science of all its mystery, and shows us what its
power really is. With respect to specific results it
yields us nothing that we could not reach in a sufficiently long time without methods. There is no problem in all mathematics that cannot be solved by direct
counting. But with the present implements of mathe-

matics many operations of counting can be performed in a few minutes which without mathematical methods would take a lifetime.   Just as a single human being, restricted wholly to the fruits of his own labor, could never amass a fortune, but on the contrary the accumulation of the labor of many men in the hands of one is the foundation of wealth and power, so, also, no knowledge worthy of the name can be gathered up in a single human mind limited to the span of a human life and gifted only with finite powers, except by the most exquisite economy of thought and by the careful amassment of the economically ordered experience of thousands of co-workers.   What strikes us here as the fruits of sorcery are simply the rewards of excellent housekeeping, as are the like results in civil life.   But the business of science has this advantage over every other enterprise, that from *its* amassment of wealth no one suffers the least loss.   This, too, is its blessing, its freeing and saving power.

The recognition of the economical character of science will now help us, perhaps, to understand better certain physical notions.

Those elements of an event which we call " cause and effect " are certain salient features of it, which are important for its mental reproduction.   Their importance wanes and the attention is transferred to fresh characters the moment the event or experience in question becomes familiar.   If the connexion of such features strikes us as a necessary one, it is simply be-

cause the interpolation of certain intermediate links
with which we are very familiar, and which possess,
therefore, higher authority for us, is often attended
with success in our explanations. That *ready* experience
fixed in the mosaic of the mind with which we meet
new events, Kant calls an innate concept of the understanding (*Verstandesbegriff*).

The grandest principles of physics, resolved into
their elements, differ in no wise from the descriptive
principles of the natural historian. The question,
"Why?" which is always appropriate where the explanation of a contradiction is concerned, like all proper
habitudes of thought, can overreach itself and be asked
where nothing remains to be understood. Suppose we
were to attribute to nature the property of producing
like effects in like circumstances; just these like circumstances we should not know how to find. Nature
exists once only. Our schematic mental imitation alone
produces like events. Only in the mind, therefore, does
the mutual dependence of certain features exist.

All our efforts to mirror the world in thought would
be futile if we found nothing permanent in the varied
changes of things. It is this that impels us to form the
notion of substance, the source of which is not different from that of the modern ideas relative to the conservation of energy. The history of physics furnishes
numerous examples of this impulse in almost all fields,
and pretty examples of it may be traced back to the
nursery. "Where does the light go to when it is put

out? " asks the child.   The sudden shrivelling up of a hydrogen balloon is inexplicable to a child; it looks everywhere for the large body which was just there but is now gone.

Where does heat come from ?   Where does heat go to ?   Such childish questions in the mouths of mature men shape the character of a century.

In mentally separating a body from the changeable environment in which it moves, what we really do is to extricate a group of sensations on which our thoughts are fastened and which is of relatively greater stability than the others, from the stream of all our sensations.   Absolutely unalterable this group is not. Now this, now that member of it appears and disappears, or is altered.   In its full identity it never recurs.   Yet the sum of its constant elements as compared with the sum of its changeable ones, especially if we consider the continuous character of the transition, is always so great that for the purpose in hand the former usually appear sufficient to determine the body's identity.   But because we can separate from the group every single member without the body's ceasing to be for us the same, we are easily led to believe that after abstracting all the members something additional would remain.   It thus comes to pass that we form the notion of a substance distinct from its attributes, of a thing-in-itself, whilst our sensations are regarded merely as symbols or indications of the properties of this thing-in-itself.   But it would be much better to

say that bodies or things are compendious mental symbols for groups of sensations—symbols that do not exist outside of thought. Thus, the merchant regards
the labels of his boxes merely as indexes of their contents, and not the contrary. He invests their contents, not their labels, with real value. The same
economy which induces us to analyse a group and to
establish special signs for its component parts, parts
which also go to make up other groups, may likewise
induce us to mark out by some single symbol a whole
group.

On the old Egyptian monuments we see objects
represented which do not reproduce a single visual
impression, but are composed of various impressions.
The heads and the legs of the figures appear in profile, the head-dress and the breast are seen from the
front, and so on. We have here, so to speak, a mean
view of the objects, in forming which the sculptor has
retained what he deemed essential, and neglected what
he thought indifferent. We have living exemplifica-
tions of the processes put into stone on the walls of
these old temples, in the drawings of our children, and
we also observe a faithful analogue of them in the formation of ideas in our own minds. Only in virtue of
some such facility of view as that indicated, are we
allowed to speak of *a* body. When we speak of a cube
with trimmed corners—a figure which is not a cube—
we do so from a natural instinct of economy, which
prefers to add to an old familiar conception a correc-

tion instead of forming an entirely new one.   This is
the process of all judgment.

   The crude notion of "body" can no more stand
the test of analysis than can the art of the Egyptians
or that of our little children.   The physicist who sees
a body flexed, stretched, melted, and vaporised, cuts
up this body into smaller permanent parts ; the chem-
ist splits it up into elements.   Yet even an element is
not unalterable.   Take sodium.   When warmed, the
white, silvery mass becomes a liquid, which, when the
heat is increased and the air shut out, is transformed
into a violet vapor, and on the heat being still more
increased glows with a yellow light.   If the name sodium is still retained, it is because of the continuous
character of the transitions and from a necessary instinct of economy.   By condensing the vapor, the
white metal may be made to reappear.   Indeed, even
after the metal is thrown into water and has passed
into sodium hydroxide, the vanished properties may
by skilful treatment still be made to appear ; just as a
moving body which has passed behind a column and
is lost to view for a moment may make its appearance
after a time.   It is unquestionably very convenient
always to have ready the name and thought for a
group of properties wherever that group by any possibility can appear.   But more than a compendious economical symbol for these phenomena, that name and
thought is not.   It would be a mere empty word for
one in whom it did not awaken a large group of well-

ordered sense-impressions. And the same is true of
the molecules and atoms into which the chemical element is still further analysed.

True, it is customary to regard the conservation of
weight, or, more precisely, the conservation of mass,
as a direct proof of the constancy of matter. But this
proof is dissolved, when we go to the bottom of it,
into such a multitude of instrumental and intellectual
operations, that in a sense it will be found to constitute simply an equation which our ideas in imitating
facts have to satisfy. That obscure, mysterious lump
which we involuntarily add in thought, we seek for in
vain outside the mind.

It is always, thus, the crude notion of substance
that is slipping unnoticed into science, proving itself
constantly insufficient, and ever under the necessity of
being reduced to smaller and smaller world-particles.
Here, as elsewhere, the lower stage is not rendered
indispensable by the higher which is built upon it, no
more than the simplest mode of locomotion, walking,
is rendered superfluous by the most elaborate means of
transportation. Body, as a compound of light and
touch sensations, knit together by sensations of space,
must be as familiar to the physicist who seeks it, as to
the animal who hunts its prey. But the student of the
theory of knowledge, like the geologist and the astronomer, must be permitted to reason back from the forms
which are created before his eyes to others which he
finds ready made for him.

All physical ideas and principles are succinct directions, frequently involving subordinate directions, for the employment of economically classified experiences, ready for use.   Their conciseness, as also the fact that their contents are rarely exhibited in full, often invests them with the semblance of independent existence.   Poetical myths regarding such ideas,—for example, that of Time, the producer and devourer of all things,—do not concern us here.   We need only remind the reader that even Newton speaks of an *absolute* time independent of all phenomena, and of an absolute space—views which even Kant did not shake off, and which are often seriously entertained to-day. For the natural inquirer, determinations of time are merely abbreviated statements of the dependence of one event upon another, and nothing more.   When we say the acceleration of a freely falling body is $9 \cdot 810$ metres per second, we mean the velocity of the body with respect to the centre of the earth is $9 \cdot 810$ metres greater when the earth has performed an additional 8640oth part of its rotation—a fact which itself can be determined only by the earth's relation to other heavenly bodies.   Again, in velocity is contained simply a relation of the position of a body to the position of the earth.*   Instead of referring events to the earth we may refer them to a clock, or even to our internal sensation of time.   Now, because all are connected,

* It is clear from this that all so-called elementary (differential) laws involve a relation to the Whole.

and each may be made the measure of the rest, the illusion easily arises that time has significance independently of all.*

The aim of research is the discovery of the equations which subsist between the elements of phenomena. The equation of an ellipse expresses the universal *conceivable* relation between its co-ordinates, of which only the real values have *geometrical* significance. Similarly, the equations between the elements of *phenomena* express a universal, mathematically conceivable relation. Here, however, for many values only certain directions of change are *physically* admissible. As in the ellipse only certain *values* satisfying the equation are realised, so in the physical world only certain *changes* of value occur. Bodies are always accelerated towards the earth. Differences of temperature, left to themselves, always grow less ; and so on. Similarly, with respect to space, mathematical and physiological researches have shown that the space of experience is simply an *actual* case of many conceivable cases, about whose peculiar properties experience alone can instruct us. The elucidation which this idea diffuses cannot be questioned, despite the absurd uses to which it has been put.

Let us endeavor now to summarise the results of

* If it be objected, that in the case of perturbations of the velocity of rotation of the earth, we could be sensible of such perturbations, and being obliged to have some measure of time, we should resort to the period of vibration of the waves of sodium light,—all that this would show is that for practical reasons we should select that event which best served us as the *simplest* common measure of the others.

our survey.   In the economical schematism of science lie both its strength and its weakness.   Facts are always represented at a sacrifice of completeness and never with greater precision than fits the needs of the moment.   The incongruence between thought and experience, therefore, will continue to subsist as long as the two pursue their course side by side ; but it will be continually diminished.

In reality, the point involved is always the completion of some partial experience ; the derivation of one portion of a phenomenon from some other.   In this act our ideas must be based directly upon sensations.   We call this measuring.*   The condition of science, both in its origin and in its application, is a *great relative stability* of our environment.   What it teaches us is interdependence.   Absolute forecasts, consequently, have no significance in science.   With great changes in celestial space we should lose our co-ordinate systems of space and time.

When a geometer wishes to understand the form of a curve, he first resolves it into small rectilinear elements.   In doing this, however, he is fully aware that these elements are only provisional and arbitrary devices for comprehending in parts what he cannot comprehend as a whole.   When the law of the curve is found he no longer thinks of the elements.   Similarly, it would not become physical science to see in its self-

---

* Measurement, in fact, is the definition of one phenomenon by another (standard) phenomenon.

created, changeable, economical tools, molecules and
atoms, realities behind phenomena, forgetful of the
lately acquired sapience of her older sister, philosophy,
in substituting a mechanical mythology for the old
animistic or metaphysical scheme, and thus creating
no end of suppositious problems.   The atom must remain a tool for representing phenomena, like the
functions of mathematics.   Gradually, however, as
the intellect, by contact with its subject-matter, grows
in discipline, physical science will give up its mosaic
play with stones and will seek out the boundaries and
forms of the bed in which the living stream of phenomena flows.   The goal which it has set itself is the
simplest and *most economical* abstract expression of facts.

*   *   *

The question now remains, whether the same
method of research which till now we have tacitly restricted to physics, is also applicable in the psychical
domain.   This question will appear superfluous to the
physical inquirer.   Our physical and psychical views
spring in exactly the same manner from instinctive
knowledge.   We read the thoughts of men in their
acts and facial expressions without knowing how.
Just as we predict the behavior of a magnetic needle
placed near a current by imagining Ampère's swimmer in the current, similarly we predict in thought the
acts and behavior of men by assuming sensations, feelings, and wills similar to our own connected with their
bodies.   What we here instinctively perform would

appear to us as one of the subtlest achievements of science, far outstripping in significance and ingenuity Ampère's rule of the swimmer, were it not that every child unconsciously accomplished it. The question simply is, therefore, to grasp scientifically, that is, by conceptional thought, what we are already familiar with from other sources. And here much is to be accomplished. A long sequence of facts is to be disclosed between the physics of expression and movement and feeling and thought.

We hear the question, "But how is it possible to explain feeling by the motions of the atoms of the brain?" Certainly this will never be done, no more than light or heat will ever be deduced from the law of refraction. We need not deplore, therefore, the lack of ingenious solutions of this question. The problem is not a problem. A child looking over the walls of a city or of a fort into the moat below sees with astonishment living people in it, and not knowing of the portal which connects the wall with the moat, cannot understand how they could have got down from the high ramparts. So it is with the notions of physics. We cannot climb up into the province of psychology by the ladder of our abstractions, but we can climb down into it.

Let us look at the matter without bias. The world consists of colors, sounds, temperatures, pressures, spaces, times, and so forth, which now we shall not call sensations, nor phenomena, because in either term an arbitrary, one-sided theory is embodied, but simply *elements.* The fixing of the flux of these elements, whether mediately or immediately, is the real object of physical research. As long as, neglecting our own body, we employ ourselves with the interdependence of those groups of elements which, including men and animals, make up *foreign* bodies, we are physicists. For example, we investigate the change of the red color of a body as produced by a change of illumination. But the moment we consider the special influence on the red of the elements constituting our body, outlined by the well-known perspective with head invisible, we are at work in the domain of physiological psychology. We close our eyes, and the red together with the whole visible world disappears. There exists, thus, in the perspective field of every sense a portion which exercises on all the rest a different and more powerful influence than the rest upon one another. With this, however, all is said. In the light of this remark, we call *all* elements, in so far as we regard them as dependent on this special part (our body), *sensations.* That the world is our sensation, in this sense, cannot be questioned. But to make a system of conduct out of this provisional conception, and to abide its slaves, is as unnecessary for us as would be a similar course for a mathematician who, in varying a series of variables of a function which were previously assumed to be constant, or in interchanging the independent variables, finds his method to be the source of some very surprising ideas for him.*

If we look at the matter in this unbiassed light it will appear indubitable that the method of physiological psychology is none other than that of physics; what is more, that this science is a part of physics. Its subject-matter is not different from that of physics. It will unquestionably determine the relations the sensations bear to the physics of our body. We have already learned from a member of this academy (Hering) that in all probability a sixfold manifoldness of the chemical processes of the visual substance corresponds to the sixfold manifoldness of color-sensation, and a threefold manifoldness of the physiological processes to the threefold manifoldness of space-sensations. The paths of reflex actions and of the will are followed up and disclosed; it is ascertained what region of the brain subserves the function of speech, what region the function of locomotion, etc. That which still clings to our body, namely, our thoughts, will, when those investigations are finished, present no difficulties new in principle. When experience has once clearly exhibited these facts and science has

* I have represented the point of view here taken for more than thirty years and developed it in various writings (*Erhaltung der Arbeit*, 1872, parts of which are published in the article on *The Conservation of Energy* in this collection; *The Forms of Liquids*, 1872, also published in this collection; and the *Bewegungsempfindungen*, 1875). The idea, though known to philosophers, is unfamiliar to the majority of physicists. It is a matter of deep regret to me, therefore, that the title and author of a small tract which accorded with my views in numerous details and which I remember having caught a glance of in a very busy period (1879–1880), have so completely disappeared from my memory that all efforts to obtain a clue to them have hitherto been fruitless.

marshalled them in economic and perspicuous order, there is no doubt that we shall *understand* them. For other "understanding" than a mental mastery of facts never existed. Science does not create facts from facts, but simply *orders* known facts.

Let us look, now, a little more closely into the modes of research of physiological psychology. We have a very clear idea of how a body moves in the space encompassing it. With our optical field of sight we are very familiar. But we are unable to state, as a rule, how we have come by an idea, from what corner of our intellectual field of sight it has entered, or by what region the impulse to a motion is sent forth. Moreover, we shall never get acquainted with this mental field of view from self-observation alone. Self-observation, in conjunction with physiological research, which seeks out physical connexions, can put this field of vision in a clear light before us, and will thus first really reveal to us our inner man.

Primarily, natural science, or physics, in its widest sense, makes us acquainted with only the firmest connexions of groups of elements. Provisorily, we may not bestow too much attention on the single constituents of those groups, if we are desirous of retaining a comprehensible whole. Instead of equations between the primitive variables, physics gives us, as much the easiest course, equations between *functions* of those variables. Physiological psychology teaches us how to separate the visible, the tangible, and the audible

from bodies—a labor which is subsequently richly re-
quited, as the division of the subjects of physics well
shows. Physiology further analyses the visible into
light and space sensations ; the first into colors, the
last also into their component parts ; it resolves noises
into sounds, these into tones, and so on. Unquestionably this analysis can be carried much further than it
has been. It will be possible in the end to exhibit the
common elements at the basis of very abstract but
definite logical acts of like form,—elements which the
acute jurist and mathematician, as it were, *feels* out,
with absolute certainty, where the uninitiated hears
only empty words. Physiology, in a word, will reveal
to us the true real elements of the world. Physiological
psychology bears to physics in its widest sense a relation similar to that which chemistry bears to physics
in its narrowest sense. But far greater than the mutual support of physics and chemistry will be that
which natural science and psychology will render each
other. And the results that shall spring from this
union will, in all likelihood, far outstrip those of the
modern mechanical physics.

What those ideas are with which we shall comprehend the world when the closed circuit of physical and
psychological facts shall lie complete before us, (that
circuit of which we see now only two disjoined parts,)
cannot be foreseen at the outset of the work. The
men will be found who will see what is right and
will have the courage, instead of wandering in the

intricate paths of logical and historical accident, to
enter on the straight ways to the heights from which
the mighty stream of facts can be surveyed. Whether
the notion which we now call matter will continue to
have a scientific significance beyond the crude purposes of common life, we do not know. But we certainly shall wonder how colors and tones which were
such innermost parts of us could suddenly get lost in
our physical world of atoms; how we could be suddenly surprised that something which outside us simply clicked and beat, in our heads should make light
and music; and how we could ask whether matter can
feel, that is to say, whether a mental symbol for a
group of sensations can feel?

We cannot mark out in hard and fast lines the
science of the future, but we can foresee that the rigid
walls which now divide man from the world will gradually disappear; that human beings will not only confront each other, but also the entire organic and so-
called lifeless world, with less selfishness and with live-
lier sympathy. Just such a presentiment as this perhaps possessed the great Chinese philosopher Licius
some two thousand years ago when, pointing to a heap
of mouldering human bones, he said to his scholars in
the rigid, lapidary style of his tongue: "These and I
alone have the knowledge that we neither live nor are
dead."

ON TRANSFORMATION AND ADAPTA-
TION IN SCIENTIFIC THOUGHT.*

IT was towards the close of the sixteenth century that Galileo with a superb indifference to the dialectic arts and sophistic subtleties of the Schoolmen of his time, turned the attention of his brilliant mind to nature. By nature his ideas were transformed and released from the fetters of inherited prejudice. At once the mighty revolution was felt, that was therewith effected in the realm of human thought—felt indeed in circles far remote and wholly unrelated to the sphere of science, felt in strata of society that hitherto had only indirectly recognised the influence of scientific thought.

* Inaugural Address, delivered on assuming the Rectorate of the University of Prague, October 18, 1883.

The idea presented in this essay is neither new nor remote. I have touched upon it myself on several occasions (first in 1867), but have never made it the subject of a formal disquisition. Doubtless, others, too, have treated it; it lies, so to speak, in the air. However, as many of my illustrations were well received, although known only in an imperfect form from the lecture itself and the newspapers, I have, contrary to my original intention, decided to publish it. It is not my intention to trespass here upon the domain of biology. My statements are to be taken merely as the expression of the fact that no one can escape the influence of a great and far-reaching idea.

And how great and how far-reaching that revolution was! From the beginning of the seventeenth century till its close we see arising, at least in embryo, almost all that plays a part in the natural and technical science of to-day, almost all that in the two centuries following so wonderfully transformed the facial appearance of the earth, and all that is moving onward in process of such mighty evolution to-day. And all this, the direct result of Galilean ideas, the direct outcome of that freshly awakened sense for the investigation of natural phenomena which taught the Tuscan philosopher to form the concept and the law of falling bodies from the *observation* of a falling stone! Galileo began his investigations without an implement worthy of the name; he measured time in the most primitive way, by the efflux of water. Yet soon afterwards the telescope, the microscope, the barometer, the thermometer, the air-pump, the steam-engine, the pendulum, and the electrical machine were invented in rapid succession. The fundamental theorems of dynamical science, of optics, of heat, and of electricity were all disclosed in the century that followed Galileo.

Of scarcely less importance, it seems, was that movement which was prepared for by the illustrious biologists of the hundred years just past, and formally begun by the late Mr. Darwin. Galileo quickened the sense for the simpler phenomena of *inorganic* nature. And with the same simplicity and frankness that marked the efforts of Galileo, and without the aid of technical or scientific instruments, without physical or chemical experiment, but solely by the power of thought and observation, Darwin grasps a new property of *organic.* nature—which we may briefly call its *plasticity.** With the same directness of purpose, Darwin, too, pursues his way. With the same candor and love of truth, he points out the strength and the weakness of his demonstrations. With masterly equanimity he holds aloof from the discussion of irrelevant subjects and wins alike the admiration of his adherents and of his adversaries.

Scarcely thirty years have elapsed† since Darwin first propounded the principles of his theory of evolution.

*At first sight an apparent contradiction arises from the admission of both heredity and adaptation ; and it is undoubtedly true that a strong disposition to heredity precludes great capability of adaptation. But imagine the organism to be a plastic mass which retains the form transmitted to it by former influences until new influences modify it ; the *one* property of *plasticity* will then represent capability of adaptation as well as power of heredity. Analogous to this is the case of a bar of magnetised steel of high coercive force: the steel retains its magnetic properties until a new force displaces them. Take also a body in motion : the body retains the velocity acquired in (*inherited* from) the interval of time just preceding, except it be changed in the next moment by an accelerating force. In the case of the body in motion the *change* of velocity (*Abänderung*) was looked upon as a matter of course, while the discovery of the principle of *inertia* (or persistence) created surprise ; in Darwin's case, on the contrary, *heredity* (or persistence) was taken for granted, while the principle of *variation* (*Abänderung*) appeared novel.

Fully adequate views are, of course, to be reached only by a study of the original facts emphasised by Darwin, and not by these analogies. The example referring to motion, if I am not mistaken, I first heard, in conversation, from my friend J. Popper, Esq., of Vienna.

Many inquirers look upon the stability of the species as something settled, and oppose to it the Darwinian theory. But the stability of the species is itself a "theory." The essential modifications which Darwin's views also are undergoing will be seen from the works of Wallace [and Weismann], but more especially from a book of W. H. Rolph, *Biologische Probleme*, Leipsic, 1882. Unfortunately, this last talented investigator is no longer numbered among the living.

† Written in 1883.

Yet, already we see his ideas firmly rooted in every branch of human thought, however remote. Everywhere, in history, in philosophy, even in the physical sciences, we hear the watchwords: heredity, adaptation, selection. We speak of the struggle for existence among the heavenly bodies and of the struggle for existence in the world of molecules.*

The impetus given by Galileo to scientific thought was marked in every direction; thus, his pupil, Borelli, founded the school of exact medicine, from whence proceeded even distinguished mathematicians. And now Darwinian ideas, in the same way, are animating all provinces of research. It is true, nature is not made up of two distinct parts, the inorganic and the organic; nor must these two divisions be treated perforce by totally distinct methods. Many *sides*, however, nature has. Nature is like a thread in an intricate tangle, which must be followed and traced, now from this point, now from that. But we must never imagine, —and this physicists have learned from Faraday and J. R. Mayer,—that progress along paths once entered upon is the *only* means of reaching the truth.

It will devolve upon the specialists of the future to determine the relative tenability and fruitfulness of the Darwinian ideas in the different provinces. Here I wish simply to consider the growth of natural *knowledge* in the light of the theory of evolution. For knowledge, too, is a product of organic nature. And although

* See Pfaundler, *Pogg. Ann.*, *Jubelband*, p. 182.

ideas, as such, do not comport themselves in all respects like independent organic individuals, and although violent comparisons should be avoided, still, if Darwin reasoned rightly, the general imprint of evolution and transformation must be noticeable in ideas also.

I shall waive here the consideration of the fruitful topic of the transmission of ideas or rather of the transmission of the aptitude for certain ideas.* Nor would it come within my province to discuss psychical evolution in any form, as Spencer† and many other modern psychologists have done, with varying success. Neither shall I enter upon a discussion of the struggle for existence and of natural selection among scientific theories.‡ We shall consider here only such processes of transformation as every student can easily observe in his own mind.

*   *   *

The child of the forest picks out and pursues with marvellous acuteness the trails of animals. He outwits and overreaches his foes with surpassing cunning. He is perfectly at home in the sphere of his peculiar experience. But confront him with an unwonted phenomenon ; place him face to face with a technical product of modern civilisation, and he will lapse into impotency and helplessness. Here are facts which he

---

* See the beautiful discussions of this point in Hering's *Memory as a General Function of Organised Matter* (1870), Chicago, The Open Court Publishing Co., 1887. Compare also Dubois, *Ueber die Uebung*, Berlin, 1881.

† Spencer, *The Principles of Psychology*. London, 1872.

‡ See the article *The Velocity of Light*, page 63.

does not comprehend. If he endeavors to grasp their meaning, he misinterprets them. He fancies the moon, when eclipsed, to be tormented by an evil spirit. To his mind a puffing locomotive is a living monster. The letter accompanying a commission with which he is entrusted, having once revealed his thievishness, is in his imagination a conscious being, which he must hide beneath a stone, before venturing to commit a fresh trespass. Arithmetic to him is like the art of the geomancers in the Arabian Nights,—an art which is able to accomplish every imaginable impossibility. And, like Voltaire's *ingénu*, when placed in our social world, he plays, as we think, the maddest pranks.

With the man who has made the achievements of modern science and civilisation his own, the case is quite different. He sees the moon pass temporarily into the shadow of the earth. He feels in his thoughts the water growing hot in the boiler of the locomotive; he feels also the increase of the tension which pushes the piston forward. Where he is not able to trace the direct relation of things he has recourse to his yard-stick and table of logarithms, which aid and facilitate his thought without predominating over it. Such opinions as he cannot concur in, are at least known to him, and he knows how to meet them in argument.

Now, wherein does the difference between these two men consist? The train of thought habitually employed by the first one does not correspond to the facts that he sees. He is surprised and nonplussed

at every step.   But the thoughts of the second man
follow and anticipate events, his thoughts have become adapted or accommodated to the larger field of
observation and activity in which he is located; he conceives things as they are.   The Indian's sphere of experience, however, is quite different; his bodily organs
of sense are in constant activity; he is ever intensely
alert and on the watch for his foes; or, his entire attention and energy are engaged in procuring suste-
nance.   Now, how can such a creature project his mind
into futurity, foresee or prophesy?   This is not possible until our fellow-beings have, in a measure, relieved
us of our concern for existence.   It is then that we
acquire freedom for observation, and not infrequently
too that narrowness of thought which society helps and
teaches us to disregard.

   If we move for a time within a fixed circle of phenomena which recur with unvarying uniformity, our
thoughts gradually adapt themselves to our environment; our ideas reflect unconsciously our surroundings.   The stone we hold in our hand, when dropped,
not only falls to the ground in reality; it also falls in
our thoughts.   Iron-filings dart towards a magnet in
imagination as well as in fact, and, when thrown into
a fire, they grew hot in conception as well.

   The impulse to complete mentally a phenomenon
that has been only partially observed, has not its origin
in the phenomenon itself; of this fact, we are fully
sensible.   And we well know that it does not lie within

the sphere of our volition. It seems to confront us rather as a power and a law imposed from without and controlling both thought and facts.

The fact that we are able by the help of this law to prophesy and forecast, merely proves a sameness or uniformity of environment sufficient to effect a mental adaptation of this kind. A necessity of fulfilment, however, is not contained in this compulsory principle which controls our thoughts; nor is it in any way determined by the possibility of prediction. We are always obliged, in fact, to await the completion of what has been predicted. Errors and departures are constantly discernible, and are slight only in provinces of great rigid constancy, as in astronomy.

In cases where our thoughts follow the connexion of events with ease, and in instances where we positively forefeel the course of a phenomenon, it is natural to fancy that the latter is determined by and must conform to our thoughts. But the belief in that mysterious agency called *causality*, which holds thought and event in unison, is violently shaken when a person first enters a province of inquiry in which he has previously had no experience. Take for instance the strange interaction of electric currents and magnets, or the reciprocal action of currents, which seem to defy all the resources of mechanical science. Let him be confronted with such phenomena and he will immediately feel himself forsaken by his power of prediction; he will bring nothing with him into this strange field of events but the hope of soon being able to adapt his ideas to the new conditions there presented.

A person constructs from a bone the remaining anatomy of an animal; or from the visible part of a half-concealed wing of a butterfly he infers and reconstructs the part concealed. He does so with a feeling of highest confidence in the accuracy of his results; and in these processes we find nothing preternatural or transcendent. But when physicists adapt their thoughts to conform to the dynamical course of events in time, we invariably surround their investigations with a metaphysical halo; yet these latter adaptations bear quite the same character as the former, and our only reason for investing them with a metaphysical garb, perhaps, is their high practical value.*

Let us consider for a moment what takes place when the field of observation to which our ideas have been adapted and now conform, becomes enlarged. We had, let us say, always seen heavy bodies sink when their support was taken away; we had also seen, perhaps, that the sinking of heavier bodies forced lighter bodies upwards. But now we see a lever in action, and we are suddenly struck with the fact that a lighter body is lifting another of much greater weight.

* I am well aware that the endeavor to confine oneself in natural research to *facts* is often censured as an exaggerated fear of metaphysical spooks. But I would observe, that, judged by the mischief which they have wrought, the metaphysical, of all spooks, are the least fabulous. It is not to be denied that many forms of thought were not originally acquired by the individual, but were antecedently formed, or rather prepared for, in the development of the species, in some such way as Spencer, Haeckel, Hering, and others have supposed, and as I myself have hinted on various occasions.

Our customary train of thought demands its rights; the new and unwonted event likewise demands its rights. From this conflict between thought and fact the *problem* arises; out of this partial contrariety springs the question, "Why?" With the new adaptation to the enlarged field of observation, the problem disappears, or, in other words, is solved. In the instance cited, we must adopt the habit of always considering the mechanical work performed.

The child just awakening into consciousness of the world, knows no problem. The bright flower, the ringing bell, are all new to it; yet it is surprised at nothing. The out and out Philistine, whose only thoughts lie in the beaten path of his every-day pursuits, likewise has no problems. Everything goes its wonted course, and if perchance a thing go wrong at times, it is at most a mere object of curiosity and not worth serious consideration. In fact, the question "Why?" loses all warrant in relations where we are familiar with every aspect of events. But the capable and talented young man has his head full of problems; he has acquired, to a greater or less degree, certain habitudes of thought, and at the same time he is constantly observing what is new and unwonted, and in his case there is no end to the questions, "Why?"

Thus, the factor which most promotes scientific thought is the gradual widening of the field of experience. We scarcely notice events we are accustomed to; the latter do not really develop their intellectual

significance until placed in contrast with something to
which we are unaccustomed.   Things that at home
are passed by unnoticed, delight us when abroad,
though they may appear in only slightly different forms.
The sun shines with heightened radiance, the flowers
bloom in brighter colors, our fellow-men accost us
with lighter and happier looks.   And, returning home,
we find even the old familiar scenes more inspiring
and suggestive than before.

Every motive that prompts and stimulates us to
modify and transform our thoughts, proceeds from
what is new, uncommon, and not understood. Novelty
excites wonder in persons whose fixed habits of thought
are shaken and disarranged by what they see. But the
element of wonder never lies in the phenomenon or
event observed; its place is in the person observing.
People of more vigorous mental type aim at once at an
adaptation of thought that will conform to what they
have observed.   Thus does science eventually become
the natural foe of the wonderful.   The sources of the
marvellous are unveiled, and surprise gives way to
calm interpretation.

Let us consider such a mental transformative process in detail.   The circumstance that heavy bodies
fall to the earth appears perfectly natural and regular.
But when a person observes that wood floats upon
water, and that flames and smoke rise in the air, then
the contrary of the first phenomenon is presented.
An olden theory endeavors to explain these facts by im-

puting to substances the power of volition, as that attribute which is most familiar to man. It asserted that every substance seeks its proper place, heavy bodies tending downwards and light ones upwards. It soon turned out, however, that even smoke had weight, that it, too, sought its place below, and that it was forced upwards only because of the downward tendency of the air, as wood is forced to the surface of water because the water exerts the greater downward pressure.

Again, we see a body thrown into the air. It ascends. How is it that it does not seek its proper place? Why does the velocity of its "violent" motion decrease as it rises, while that of its "natural" fall increases as it descends. If we mark closely the relation between these two facts, the problem will solve itself. We shall see, as Galileo did, that the decrease of velocity in rising and the increase of velocity in falling are one and the same phenomenon, viz., an increase of velocity towards the earth. Accordingly, it is not a place that is assigned to the body, but an increase of velocity towards the earth.

By this idea the movements of heavy bodies are rendered perfectly familiar. Newton, now, firmly grasping this new way of thinking, sees the moon and the planets moving in their paths upon principles similar to those which determine the motion of a projectile thrown into the air. Yet the movements of the planets were marked by peculiarities which compelled

him once more to modify slightly his customary mode
of thought.   The heavenly bodies, or rather the parts
composing them, do not move with constant accelerations towards each other, but ''attract each other,''
directly as the mass and inversely as the square of the
distance.

This latter notion, which includes the one applying
to terrestrial bodies as a special case, is, as we see,
quite different from the conception from which we
started.   How limited in scope was the original idea
and to what a multitude of phenomena is not the present one applicable !   Yet there is a trace, after all,
of the ''search for place '' in the expression ''attraction.''   And it would be folly, indeed, for us to avoid,
with punctilious dread, this conception of ''attraction''
as bearing marks of its pedigree.   It is the historical
base of the Newtonian conception and it still continues
to direct our thoughts in the paths so long familiar to
us.   Thus, the happiest ideas do not fall from heaven,
but spring from notions already existing.

Similarly, a ray of light was first regarded as a continuous and homogeneous straight line.   It then became the path of projection for minute missiles; then
an aggregate of the paths of countless different kinds
of missiles.   It became periodic ; it acquired various
sides ; and ultimately it even lost its motion in a
straight line.

The electric current was conceived originally as
the flow of a hypothetical fluid.   To this conception

was soon added the notion of a chemical current, the notion of an electric, magnetic, and anisotropic optical field, intimately connected with the path of the current. And the richer a conception becomes in following and keeping pace with facts, the better adapted it is to anticipate them.

Adaptive processes of this kind have no assignable beginning, inasmuch as every problem that incites to new adaptation, presupposes a fixed habitude of thought. Moreover, they have no visible end ; in so far as experience never ceases. Science, accordingly, stands midway in the evolutionary process ; and science may advantageously direct and promote this process, but it can never take its place. That science is inconceivable the principles of which would enable a person with no experience to construct the world of experience, without a knowledge of it. One might just as well expect to become a great musician, solely by the aid of theory, and without musical experience; or to become a painter by following the directions of a text-book.

In glancing over the history of an idea with which we have become perfectly familiar, we are no longer able to appreciate the full significance of its growth. The deep and vital changes that have been effected in the course of its evolution, are recognisable only from the astounding narrowness of view with which great contemporary scientists have occasionally opposed each other. Huygens's wave-theory of light was in-

comprehensible to Newton, and Newton's idea of universal gravity was unintelligible to Huygens.   But a
century afterwards both notions were reconcilable,
even in ordinary minds.

On the other hand, the original creations of pioneer intellects, unconsciously formed, do not assume
a foreign garb; their form is their own.   In them,
childlike simplicity is joined to the maturity of man-
hood, and they are not to be compared with processes
of thought in the average mind.   The latter are carried
on as are the acts of persons in the state of mesmerism,
where actions involuntarily follow the images which
the words of other persons suggest to their minds.

The ideas that have become most familiar through
long experience, are the very ones that intrude themselves into the conception of every new fact observed.
In every instance, thus, they become involved in a
struggle for self-preservation, and it is just they that
are seized by the inevitable process of transformation.

Upon this process rests substantially the method
of explaining by hypothesis new and uncomprehended
phenomena.   Thus, instead of forming entirely new
notions to explain the movements of the heavenly
bodies and the phenomena of the tides, we imagine the
material particles composing the bodies of the universe
to possess weight or gravity with respect to one another.   Similarly, we imagine electrified bodies to be
freighted with fluids that attract and repel, or we conceive the space between them to be in a state of elas-

tic tension. In so doing, we substitute for new ideas distinct and more familiar notions of old experience— notions which to a great extent run unimpeded in their courses, although they too must suffer partial transformation.

The animal cannot construct new members to perform every new function that circumstances and fate demand of it. On the contrary it is obliged to make use of those it already possesses. When a vertebrate animal chances into an environment where it must learn to fly or swim, an additional pair of extremities is not grown for the purpose. On the contrary, the animal must adapt and transform a pair that it already has.

The construction of hypotheses, therefore, is not the product of artificial scientific methods. This process is unconsciously carried on in the very infancy of science. Even later, hypotheses do not become detrimental and dangerous to progress except when more reliance is placed on them than on the facts themselves; when the contents of the former are more highly valued than the latter, and when, rigidly adhering to hypothetical notions, we overestimate the ideas we possess as compared with those we have to acquire.

The extension of our sphere of experience always involves a transformation of our ideas. It matters not whether the face of nature becomes actually altered, presenting new and strange phenomena, or whether these phenomena are brought to light by an intentional or accidental turn of observation.   In fact, all the varied methods of scientific inquiry and of purposive mental adaptation enumerated by John Stuart Mill, those of observation as well as those of experiment, are ultimately recognisable as forms of one fundamental method, the method of change, or variation.   It is through change of circumstances that the natural philosopher learns. This process, however, is by no means confined to the investigator of nature. The historian, the philosopher, the jurist, the mathematician, the artist, the æsthetician,* all illuminate and unfold their ideas by producing from the rich treasures of memory similar, but different, cases ; thus, they observe and experiment in their thoughts.   Even if all sense-experience should suddenly cease, the events of the days past would meet in different attitudes in the mind and the process of adaptation would still continue—a process which, in contradistinction to the adaptation of thoughts to facts in practical spheres, would be strictly theoretical, being an adaptation of thoughts to thoughts.

The method of change or variation brings before us like cases of phenomena, having partly the same and partly different elements. It is only by comparing different cases of refracted light at changing angles of incidence that the common factor, the constancy of

---

* Compare, for example, Schiller, *Zerstreute Betrachtungen über verschiedene ästhetische Gegenstände.*

the refractive index, is disclosed. And only by comparing the refractions of light of different colors, does
the difference, the inequality of the indices of refraction, arrest the attention. Comparison based upon
change leads the mind simultaneously to the highest
abstractions and to the finest distinctions.

Undoubtedly, the animal also is able to distinguish
between the similar and dissimilar of two cases. Its
consciousness is aroused by a noise or a rustling, and
its motor centre is put in readiness. The sight of the
creature causing the disturbance, will, according to its
size, provoke flight or prompt pursuit; and in the latter case, the more exact distinctions will determine the
mode of attack. But man alone attains to the faculty
of voluntary and conscious comparison. Man alone
can, by his power of abstraction, rise, in one moment,
to the comprehension of principles like the conservation of mass or the conservation of energy, and in the
next observe and mark the arrangement of the iron
lines in the spectrum. In thus dealing with the objects of his conceptual life, his ideas unfold and ex-
pand, like his nervous system, into a widely ramified
and organically articulated tree, on which he may follow every limb to its farthermost branches, and, when
occasion demands, return to the trunk from which he
started.

The English philosopher Whewell has remarked
that two things are requisite to the formation of science: facts and ideas. Ideas alone lead to empty

speculation; mere facts can yield no organic knowledge.   We see that all depends upon the capacity of
adapting existing notions to fresh facts.

Over-readiness to yield to every new fact prevents
fixed habits of thought from arising.   Excessively rigid
habits of thought impede freedom of observation.   In
the struggle, in the compromise between judgment
and prejudgment (prejudice), if we may use the term,
our understanding of things broadens.

Habitual judgment, applied to a new case without
antecedent tests, we call prejudgment or prejudice.
Who does not know its terrible power! But we think
less often of the importance and utility of prejudice.
Physically, no one could exist, if he had to guide and
regulate the circulation, respiration, and digestion of
his body by conscious and purposive acts.   So, too,
no one could exist intellectually if he had to form judgments on every passing experience, instead of allow-
ing himself to be controlled by the judgments he has
already formed.   Prejudice is a sort of reflex motion
in the province of intelligence.

On prejudices, that is, on habitual judgments not
tested in every case to which they are applied, reposes
a goodly portion of the thought and work of the natural scientist.   On prejudices reposes most of the conduct of society.   With the sudden disappearance of
prejudice society would hopelessly dissolve.   That
prince displayed a deep insight into the power of intellectual habit, who quelled the loud menaces and

demands of his body-guard for arrears of pay and compelled them to turn about and march, by simply pronouncing the regular word of command ; he well knew that they would be unable to resist that.

Not until the discrepancy between habitual judgments and facts becomes great is the investigator implicated in appreciable illusion. Then tragic complications and catastrophes occur in the practical life of individuals and nations—crises where man, placing custom above life, instead of pressing it into the service of life, becomes the victim of his error. The very power which in intellectual life advances, fosters, and sustains us, may in other circumstances delude and destroy us.

* * *

Ideas are not all of life. They are only momentary efflorescences of light, designed to illuminate the paths of the will. But as delicate reagents on our organic evolution our ideas are of paramount importance. No theory can gainsay the vital transformation which we feel taking place within us through their agency. Nor is it necessary that we should have a proof of this process. We are immediately assured of it.

The transformation of ideas thus appears as a part of the general evolution of life, as a part of its adaptation to a constantly widening sphere of action. A granite boulder on a mountain-side tends towards the earth below. It must abide in its resting-place for thousands of years before its support gives way. The

shrub that grows at its base is farther advanced ; it
accommodates itself to summer and winter.   The fox
which, overcoming the force of gravity, creeps to the
summit where he has scented his prey, is freer in his
movements than either.   The arm of man reaches
further still; and scarcely anything of note happens
in Africa or Asia that does not leave an imprint upon
his life.   What an immense portion of the life of
other men is reflected in ourselves ; their joys, their
affections, their happiness and misery !   And this too,
when we survey only our immediate surroundings,
and confine our attention to modern literature.   How
much more do we experience when we travel through
ancient Egypt with Herodotus, when we stroll through
the streets of Pompeii, when we carry ourselves back
to the gloomy period of the crusades or to the golden
age of Italian art, now making the acquaintance of a
physician of Molière, and now that of a Diderot or of
a D'Alembert.   What a great part of the life of others,
of their character and their purpose, do we not absorb
through poetry and music !   And although they only
gently touch the chords of our emotions, like the memory of youth softly breathing upon the spirit of an
aged man, we have nevertheless lived them over again
in part.   How great and comprehensive does self become in this conception ; and how insignificant the
person !   Egoistical systems both of optimism and pessimism perish with their narrow standard of the import of intellectual life.   We feel that the real pearls

of life lie in the ever changing contents of consciousness, and that the person is merely an indifferent symbolical thread on which they are strung.*

We are prepared, thus, to regard ourselves and
every one of our ideas as a product and a subject of
universal evolution ; and in this way we shall advance
sturdily and unimpeded along the paths which the
future will throw open to us.†

*We must not be deceived in imagining that the happiness of other people is not a very considerable and essential portion of our own. It is common
capital, which cannot be created by the individual, and which does not perish
with him. The formal and material limitation of the *ego* is necessary and sufficient only for the crudest practical objects, and cannot subsist in a broad conception. Humanity in its entirety may be likened to a polyp-plant. The
material and organic bonds of individual union have, indeed, been severed ;
they would only have impeded freedom of movement and evolution. But the
ultimate aim, the psychical connexion of the whole, has been attained in a
much higher degree through the richer development thus made possible.

†C. E. von Baer, the subsequent opponent of Darwin and Haeckel, has
discussed in two beautiful addresses (*Das allgemeinste Gesetz der Natur in
aller Entwickelung*, and *Welche Auffassung der lebenden Natur ist die richtige, und wie ist diese Auffassung auf die Entomologie anzuwenden ?*) the
narrowness of the view which regards an animal in its existing state as
finished and complete, instead of conceiving it as a phase in the series of evolutionary forms and regarding the species itself as a phase of the development
of the animal world in general.

ON THE PRINCIPLE OF COMPARISON
IN PHYSICS.*

TWENTY years ago when Kirchhoff defined the object of mechanics as the ''description, in complete
and very simple terms, of the motions occurring in nature,'' he produced by the statement a peculiar impression. Fourteen years subsequently, Boltzmann, in the
life-like picture which he drew of the great inquirer,
could still speak of the universal astonishment at this
novel method of treating mechanics, and we meet with
epistemological treatises to-day, which plainly show
how difficult is the acceptance of this point of view. A
modest and small band of inquirers there were, however, to whom Kirchhoff's few words were tidings of a
welcome and powerful ally in the epistemological field.

Now, how does it happen that we yield our assent
so reluctantly to the philosophical opinion of an inquirer for whose scientific achievements we have only
words of praise? One reason probably is that few inquirers can find time and leisure, amid the exacting

---

*An address delivered before the General Session of the German Associa-
tion of Naturalists and Physicians, at Vienna, Sept. 24, 1894.

employments demanded for the acquisition of new
knowledge, to inquire closely into that tremendous
psychical process by which science is formed. Further,
it is inevitable that much should be put into Kirchhoff's
rigid words that they were not originally intended to
convey, and that much should be found wanting in
them that had always been regarded as an essential
element of scientific knowledge. What can mere description accomplish ? What has become of explanation, of our insight into the causal connexion of things?

*   *   *

Permit me, for a moment, to contemplate not the
results of science, but the mode of its *growth*, in a
frank and unbiassed manner. We know of only *one*
source of *immediate revelation* of scientific facts—*our
senses*. Restricted to this source alone, thrown wholly
upon his own resources, obliged to start always anew,
what could the isolated individual accomplish ? Of a
stock of knowledge so acquired the science of a distant negro hamlet in darkest Africa could hardly give
us a sufficiently humiliating conception. For there
that veritable miracle of thought-transference has already begun its work, compared with which the mir-
acles of the spiritualists are rank monstrosities—*communication by language*. Reflect, too, that by means
of the magical characters which our libraries contain
we can raise the spirits of the "the sovereign dead of
old " from Faraday to Galileo and Archimedes, through
ages of time—spirits who do not dismiss us with am-

biguous and derisive oracles, but tell us the best they know; then shall we feel what a stupendous and indispensable factor in the formation of science *communication* is. Not the dim, half-conscious *surmises* of the acute observer of nature or critic of humanity belong to science, but only that which they possess clearly enough to *communicate* to others.

But how, now, do we go about this communication of a newly acquired experience, of a newly observed fact? As the different calls and battle-cries of gregarious animals are unconsciously formed signs for a common observation or action, irrespective of the causes which produce such action—a fact that already involves the germ of the concept; so also the words of human language, which is only more highly specialised, are names or signs for universally known facts, which all can observe or have observed. If the mental representation, accordingly, follows the new fact at once and *passively*, then that new fact must, of itself, immediately be constituted and represented in thought by facts already universally known and commonly observed. Memory is always ready to put forward for *comparison* known facts which resemble the new event, or agree with it in certain features, and so renders possible that elementary internal judgment which the mature and definitively formulated judgment soon follows.

Comparison, as the fundamental condition of communication, is the most powerful inner vital element

of science. The zoölogist sees in the bones of the
wing-membranes of bats, fingers; he compares the
bones of the cranium with the vertebræ, the embryos
of different organisms with one another, and the different stages of development of the same organism
with one another. The geographer sees in Lake Garda
a fjord, in the Sea of Aral a lake in process of drying
up. The philologist compares different languages with
one another, and the formations of the same language
as well. If it is not customary to speak of comparative physics in the same sense that we speak of comparative anatomy, the reason is that in a science of
such great experimental activity the attention is turned
away too much from the *contemplative* element. But
like all other sciences, physics lives and grows by
comparison.

*   *   *

The manner in which the result of the comparison
finds expression in the communication, varies of course
very much. When we say that the colors of the spectrum are red, yellow, green, blue, and violet, the designations employed may possibly have been derived
from the technology of tattooing, or they may subsequently have acquired the significance of standing for
the colors of the rose, the lemon, the leaf, the corn-
flower, and the violet. From the frequent repetition
of such comparisons, however, made under the most
manifold circumstances, the inconstant features, as
compared with the permanent congruent features, get

so obliterated that the latter acquire a fixed significance independent of every object and connexion, or take on as we say an *abstract* or *conceptual* import. No one thinks at the word "red" of any other agreement with the rose than that of color, or at the word "straight" of any other property of a stretched cord than the sameness of direction. Just so, too, numbers, originally the names of the fingers of the hands and feet, from being used as arrangement-signs for all kinds of objects, were lifted to the plane of abstract concepts. A verbal report (communication) of a fact that uses only these purely abstract implements, we call a *direct description.*

The direct description of a fact of any great extent is an irksome task, even where the requisite notions are already completely developed. What a simplification it involves if we can say, the fact *A* now considered comports itself, not in *one*, but in *many* or in *all* its features, like an old and well-known fact *B*. The moon comports itself like a heavy body does with respect to the earth; light like a wave-motion or an electric vibration; a magnet, as if it were laden with gravitating fluids, and so on. We call such a description, in which we appeal, as it were, to a description already and elsewhere formulated, or perhaps still to be precisely formulated, an *indirect description.* We are at liberty to supplement this description, gradually, by direct description, to correct it, or to replace it altogether. We see, thus, without difficulty, that what is called a *theory* or a *theoretical idea*, falls under the category of what is here termed indirect description.

*   *   *

What, now, is a theoretical idea? Whence do we get it? What does it accomplish for us? Why does it occupy a higher place in our judgment than the mere holding fast to a fact or an observation? Here, too, memory and comparison alone are in play. But instead of *a single* feature of resemblance culled from memory, in this case *a great system* of resemblances confronts us, a well-known physiognomy, by means of which the new fact is immediately transformed into an old acquaintance. Besides, it is in the power of the idea to offer us more than we actually see in the new fact, at the first moment; it can extend the fact, and enrich it with features which we are first induced to *seek* from such suggestions, and which are often actually found. It is this *rapidity* in extending knowledge that gives to theory a preference over simple observation. But that preference is wholly *quantitative*. Qualitatively, and in real essential points, theory differs from observation neither in the mode of its origin nor in its last results.

The adoption of a theory, however, always involves a danger. For a theory puts in the place of a fact *A* in thought, always a *different*, but simpler and more familiar fact *B*, which in *some* relations can mentally represent *A*, but for the very reason that it is different, in other relations cannot represent it. If now, as may readily happen, sufficient care is not exercised, the most fruitful theory may, in special circumstances, become a downright obstacle to inquiry. Thus, the emission-theory of light, in accustoming the physicist to think of the projectile path of the "light-particles" as an undifferentiated straight-line, demonstrably impeded the discovery of the periodicity of light. By putting in the place of light the more familiar phenomena of sound, Huygens renders light in many of its features a familiar event, but with respect to polarisation, which lacks the longitudinal waves with which alone he was acquainted, it had for him a doubly strange aspect. He is unable thus to grasp in abstract thought the fact of polarisation, which is before his eyes, whilst Newton, merely by adapting to the observation his thoughts, and putting this question, "*Annon radiorum luminis diversa sunt latera?*" abstractly grasped polarisation, that is, directly described it, a century before Malus. On the other hand, if the agreement of the fact with the idea theoretically representing it, extends further than its inventor originally anticipated, then we may be led by it to unexpected discoveries, of which conical refraction, circular polarisation by total reflexion, Hertz's waves offer ready examples, in contrast to the illustrations given above.

Our insight into the conditions indicated will be improved, perhaps, by contemplating the development of some theory or other more in detail. Let us consider a magnetised bar of steel by the side of a second unmagnetised bar, in all other respects the same. The second bar gives no indication of the presence of iron-filings; the first attracts them. Also, when the iron-filings are absent, we must think of the magnetised bar as in a different condition from that of the unmag-netised. For, that the mere presence of the iron-filings does not induce the phenomenon of attraction is proved by the second unmagnetised bar. The ingenuous man, who finds in his will, as his most familiar source of power, the best facilities for comparison, conceives a species of *spirit* in the magnet. The behavior of a warm body or of an *electrified* body suggests similar ideas. This is the point of view of the oldest theory, *fetishism*, which the inquirers of the early Middle Ages had not yet overcome, and which in its last vestiges, in the conception of forces, still flourishes in modern physics. We see, thus, the *dramatic* element need no more be absent in a scientific description, than in a thrilling novel.

If, on subsequent examination, it be observed that a cold body, in contact with a hot body, warms itself, so to speak, *at the expense* of the hot body; further, that when the substances are the same, the cold body, which, let us say, has twice the mass of the other, gains only half the number of degrees of temperature that the other loses, a wholly new impression arises. The demoniac character of the event vanishes, for the supposed spirit acts not by caprice, but according to fixed laws. In its place, however, *instinctively* the

notion of a *substance* is substituted, part of which flows
over from the one body to the other, but the total
amount of which, representable by the sum of the products of the masses into the respective changes of
temperature, remains constant.   Black was the first to
be *powerfully* struck with this resemblance of thermal
processes to the motion of a substance, and under its
guidance discovered the specific heat, the heat of fusion, and the heat of vaporisation of bodies. Gaining
strength and fixity, however, from these successes,
this notion of substance subsequently stood in the way
of scientific advancement.   It blinded the eyes of the
successors of Black, and prevented them from seeing
the manifest fact, which every savage knows, that heat
is *produced* by friction.   Fruitful as that notion was
for Black, helpful as it still is to the learner to-day in
Black's special field, permanent and universal validity
as a *theory* it could never maintain. But what is essential, conceptually, in it, viz., the constancy of the pro-
duct-sum above mentioned, retains its value and may
be regarded as a *direct description* of Black's facts.

It stands to reason that those theories which push
themselves forward unsought, instinctively, and wholly
of their own accord, should have the greatest power,
should carry our thoughts most with them, and exhibit
the staunchest powers of self-preservation.   On the
other hand, it may also be observed that when critically scrutinised such theories are extremely apt to
lose their cogency.   We are constantly busied with

"substance," its modes of action have stamped themselves indelibly upon our thoughts, our vividest and clearest reminiscences are associated with it. It should cause us no surprise, therefore, that Robert Mayer and Joule, who gave the final blow to Black's substantial conception of heat, should have re-introduced the same notion of substance in a more abstract and modified form, only applying to a much more extensive field.

Here, too, the psychological circumstances which impart to the new conception its power, lie clearly before us. By the unusual redness of the venous blood in tropical climates Mayer's attention is directed to the lessened expenditure of internal heat and to the proportionately lessened *consumption of material* by the human body in those climates. But as every effort of the human organism, including its mechanical work, is connected with the consumption of material, and as work by friction can engender heat, therefore heat and work appear in kind equivalent, and between them a proportional relation must subsist. Not *every* quantity, but the appropriately calculated *sum* of the two, as connected with a proportionate consumption of material, appears *substantial*.

By exactly similar considerations, relative to the economy of the galvanic element, Joule arrived at his view; he found experimentally that the sum of the heat evolved in the circuit, of the heat consumed in the combustion of the gas developed, of the electro-mag-

netic work of the current, properly calculated, — in
short, the sum of all the effects of the battery,—is connected with a proportionate consumption of zinc. Accordingly, this sum itself has a substantial character.

Mayer was so absorbed with the view attained,
that the indestructibility of *force*, in our phraseology
work, appeared to him *a priori* evident.   "The creation or annihilation of a force," he says, "lies without the province of human thought and power." Joule
expressed himself to a similar effect : "It is manifestly
absurd to suppose that the powers with which God
has endowed matter can be destroyed." Strange to
say, on the basis of such utterances, not Joule, but
Mayer, was stamped as a metaphysician. We may
be sure, however, that both men were merely giving
expression, and that half-unconsciously, to a powerful
formal need of the new simple view, and that both
would have been extremely surprised if it had been
proposed to them that their principle should be sub-
mitted to a philosophical congress or ecclesiastical
synod for a decision upon its validity.   But with all
agreements, the attitude of these two men, in other
respects, was totally different.   Whilst Mayer represented this *formal* need with all the stupendous instinctive force of genius, we might say almost with the
ardor of fanaticism, yet was withal not wanting in the
conceptive ability to compute, prior to all other inquirers, the mechanical equivalent of heat from old
physical constants long known and at the disposal of

all, and so to set up for the new doctrine a programme
embracing all physics and physiology; Joule, on the
other hand, applied himself to the exact verification of
the doctrine by beautifully conceived and masterfully
executed experiments, extending over all departments
of physics. Soon Helmholtz too attacked the problem,
in a totally independent and characteristic manner.
After the professional virtuosity with which this physicist grasped and disposed of all the points unsettled
by Mayer's programme and more besides, what especially strikes us is the consummate critical lucidity of
this young man of twenty-six years. In his exposition
is wanting that vehemence and impetuosity which
marked Mayer's. The principle of the conservation
of energy is no self-evident or *a priori* proposition for
him. What follows, on the assumption that that proposition obtains? In this hypothetical form, he subju-
gates his matter.

I must confess, I have always marvelled at the
æsthetic and ethical taste of many of our contemporaries who have managed to fabricate out of this relation of things, odious national and personal questions,
instead of praising the good fortune that made *several*
such men work together and of rejoicing at the instructive diversity and idiosyncrasies of great minds
fraught with such rich consequences for us.

We know that still another theoretical conception
played a part in the development of the principle of
energy, which Mayer held aloof from, namely, the con-

ception that heat, as also the other physical processes, are due to motion.   But once the principle of energy has been reached, these auxiliary and transitional theories discharge no essential function, and we may regard the principle, like that which Black gave, as a contribution to the *direct description* of a widely extended domain of facts.

It would appear from such considerations not only advisable, but even necessary, with all due recognition of the helpfulness of theoretic ideas in research, yet gradually, as the new facts grow familiar, to substitute for indirect description *direct* description, which contains nothing that is unessential and restricts itself absolutely to the abstract apprehension of facts.   We might almost say, that the descriptive sciences, so called with a tincture of condescension, have, in respect of scientific character, outstripped the physical expositions lately in vogue.   Of course, a virtue has been made of necessity here.

We must admit, that it is not in our power to describe directly every fact, on the moment.   Indeed, we should succumb in utter despair if the whole wealth of facts which we come step by step to know, were presented to us all at once.   Happily, only detached and unusual features first strike us, and such we bring nearer to ourselves by *comparison* with every-day events.   Here the notions of the common speech are first developed.   The comparisons then grow more manifold and numerous, the fields of facts compared more extensive, the concepts that make direct description possible, proportionately more general and more abstract.

First we become familiar with the motion of freely falling bodies. The concepts of force, mass, and work are then carried over, with appropriate modifications, to the phenomena of electricity and magnetism. A stream of water is said to have suggested to Fourier the first distinct picture of currents of heat. A special case of vibrations of strings investigated by Taylor, cleared up for him a special case of the conduction of heat. Much in the same way that Daniel Bernoulli and Euler constructed the most diverse forms of vibrations of strings from Taylor's cases, so Fourier constructs out of simple cases of conduction the most multifarious motions of heat; and that method has extended itself over the whole of physics. Ohm forms his conception of the electric current in imitation of Fourier's. The latter, also, adopts Fick's theory of diffusion. In an analogous manner a conception of the magnetic current is developed. All sorts of stationary currents are thus made to exhibit common features, and even the condition of complete equilibrium in an extended medium shares these features with the dynamical condition of equilibrium of a stationary current. Things as remote as the magnetic lines of force of an electric current and the stream-lines of a frictionless liquid vortex enter in this way into a peculiar relationship of similarity. The con-

cept of potential, originally enunciated for a restricted province, acquires a wide-reaching applica-
bility. Things as dissimilar as pressure, temperature,
and electromotive force, now show points of agreement in relation to ideas derived by definite methods
from that concept: viz., fall of pressure, fall of temperature, fall of potential, as also with the further notions of liquid, thermal, and electric strength of current. That relationship between systems of ideas in
which the dissimilarity of every two homologous concepts as well as the agreement in logical relations
of every two homologous pairs of concepts, is clearly
brought to light, is called an *analogy*. It is an effective
means of mastering heterogeneous fields of facts in
unitary comprehension. The path is plainly shown in
which *a universal physical phenomenology* embracing all
domains, will be developed.

In the process described we attain for the first time
to what is indispensable in the direct description of
broad fields of fact—the wide-reaching *abstract concept*.
And now I must put a question smacking of the school-
master, but unavoidable : What is a concept ? Is it a
hazy representation, admitting withal of mental visualisation ? No. Mental visualisation accompanies it
only in the simplest cases, and then merely as an ad-
junct. Think, for example, of the "coefficient of self-
induction," and seek for its visualised mental image.
Or is, perhaps, the concept a mere word ? The adoption of this forlorn idea, which has been actually pro-

posed of late by a reputed mathematician would only throw us back a thousand years into the deepest scholasticism. We must, therefore, reject it.

The solution is not far to seek. We must not think that sensation, or representation, is a purely passive process. The lowest organisms respond to it with a simple reflex motion, by engulfing the prey which approaches them. In higher organisms the centripetal stimulus encounters in the nervous system obstacles and aids which modify the centrifugal process. In still higher organisms, where prey is pursued and examined, the process in question may go through extensive paths of circular motions before it comes to relative rest. Our own life, too, is enacted in such processes; all that we call science may be regarded as parts, or middle terms, of such activities.

It will not surprise us now if I say : the definition of a concept, and, when it is very familiar, even its name, is an *impulse* to some accurately determined, often complicated, critical, comparative, or constructive *activity*, the usually sense-perceptive result of which is a term or member of the concept's scope. It matters not whether the concept draws the attention only to one certain sense (as sight) or to a phase of a sense (as color, form), or is the starting-point of a complicated action; nor whether the activity in question (chemical, anatomical, and mathematical operations) is muscular or technical, or performed wholly in the imagination, or only intimated. The concept is

to the physicist what a musical note is to a piano-
player. A trained physicist or mathematician reads a
memoir like a musician reads a score. But just as the
piano-player must first learn to move his fingers singly
and collectively, before he can follow his notes without effort, so the physicist or mathematician must go
through a long apprenticeship before he gains con-
trol, so to speak, of the manifold delicate innervations
of his muscles and imagination. Think of how frequently the beginner in physics or mathematics performs more, or less, than is required, or of how frequently he conceives things differently from what they
are ! But if, after having had sufficient discipline, he
lights upon the phrase "coefficient of self-induction,"
he knows immediately what that term requires of him.
Long and thoroughly practised *actions*, which have
their origin in the necessity of comparing and representing facts by other facts, are thus the very kernel
of concepts. In fact, positive and philosophical philology both claim to have established that all roots
represent concepts and stood originally for muscular
activities alone. The slow assent of physicists to
Kirchhoff's dictum now becomes intelligible. They
best could feel the vast amount of individual labor,
theory, and skill required before the ideal of direct
description could be realised.

*   *   *

Suppose, now, the ideal of a given province of
facts is reached. Does description accomplish all that

the inquirer can ask? In my opinion, it does. Description is a building up of facts in thought, and this building up is, in the experimental sciences, often the condition of actual execution. For the physicist, to take a special case, the metrical units are the building-stones, the concepts the directions for building, and the facts the result of the building. Our mental imagery is almost a complete substitute for the fact, and by means of it we can ascertain all the fact's properties. We do not know that worst which we ourselves have made.

People require of science that it should *prophesy*, and Hertz uses that expression in his posthumous *Mechanics*. But, natural as it is, the expression is too narrow. The geologist and the palæontologist, at times the astronomer, and always the historian and the philologist, prophesy, so to speak, *backwards*. The descriptive sciences, like geometry and mathematics, prophesy neither forward or backwards, but seek from given conditions the conditioned. Let us say rather: *Science completes in thought facts that are only partly given.* This is rendered possible by description, for description presupposes the interdependence of the descriptive elements: otherwise nothing would be described.

It is said, description leaves the sense of causality unsatisfied. In fact, many imagine they understand motions better when they picture to themselves the pulling forces; and yet the *accelerations*, the facts, accomplish more, without superfluous additions. I

hope that the science of the future will discard the
idea of cause and effect, as being formally obscure ;
and in my feeling that these ideas contain a strong
tincture of fetishism, I am certainly not alone.   The
more proper course is, *to regard the abstract determinative elements of a fact as interdependent*, in a purely logical way, as the mathematician or geometer does.
True, by comparison with the will, forces are brought
nearer to our feeling; but it may be that ultimately the
will itself will be made clearer by comparison with the
accelerations of masses.

If we are asked, candidly, when is a fact *clear* to
us, we must say "when we can reproduce it by very
simple and very familiar intellectual operations, such
as the construction of accelerations, or the geometrical summation of accelerations, and so forth." The
requirement of *simplicity* is of course to the expert
a different matter from what it is to the novice. For
the first, description by a system of differential equations is sufficient ; for the second, a gradual construction out of elementary laws is required. The first
discerns at once the connexion of the two expositions.
Of course, it is not disputed that the *artistic* value of
materially equivalent descriptions may not be different.

Most difficult is it to persuade strangers that the
grand universal laws of physics, such as apply indiscriminately to material, electrical, magnetic, and other
systems, are not essentially different from descriptions.
As compared with many sciences, physics occupies in

this respect a position of vantage that is easily explained. Take, for example, anatomy. As the anatomist in his quest for agreements and differences in animals ascends to ever higher and higher *classifications*, the individual facts that represent the ultimate terms of the system, are still so different that they must be *singly* noted. Think, for example, of the common marks of the Vertebrates, of the class-characters of Mammals and Birds on the one hand and of Fishes on the other, of the double circulation of the blood on the one hand and of the single on the other. In the end, always *isolated* facts remain, which show only a *slight* likeness to one another.

A science still more closely allied to physics, chemistry, is often in the same strait. The abrupt change of the qualitative properties, in all likelihood conditioned by the slight stability of the intermediate states, the remote resemblance of the co-ordinated facts of chemistry render the treatment of its data difficult. Pairs of bodies of different qualitative properties unite in different mass-ratios; but no connexion between the first and the last is to be noted, at first.

Physics, on the other hand, reveals to us wide domains of *qualitatively homogeneous* facts, differing from one another only in the number of equal parts into which their characteristic marks are divisible, that is, differing only *quantitatively*. Even where we have to deal with qualities (colors and sounds), quantitative characters of those qualities are at our disposal. Here the classification is so simple a task that it rarely impresses us as such, whilst in infinitely fine gradations, in a *continuum of facts*, our number-system is ready beforehand to follow as far as we wish. The co-ordinated facts are here extremely similar and very closely affined, as are also their descriptions which consist in the determination of the numerical measures of one given set of characters from those of a different set by means of familiar mathematical operations—methods of derivation. Thus, the common characteristics of all descriptions can be found here ; and with them a succinct, comprehensive description, or a rule for the construction of all single descriptions, is assigned,— and this we call *law*. Well-known examples are the formulæ for freely falling bodies, for projectiles, for central motion, and so forth. If physics apparently accomplishes more by its methods than other sciences, we must remember that in a sense it has presented to it much simpler problems.

The remaining sciences, whose facts also present a physical side, need not be envious of physics for this superiority ; for all its acquisitions ultimately redound to their benefit as well. But also in other ways this mutual help shall and must change. Chemistry has advanced very far in making the methods of physics her own. Apart from older attempts, the periodical series of Lothar Meyer and Mendelejeff are a brilliant and adequate means of producing an easily surveyed system of facts, which by gradually becoming complete, will take the place almost of a continuum of facts. Further, by the study of solutions, of dissociation, in fact generally of phenomena which present a continuum of cases, the methods of thermodynamics have found entrance into chemistry. Similarly we may hope that, at some future day, a mathematician, letting the fact-continuum of embryology play before his mind, which the palæontologists of the future will supposedly have enriched with more intermediate and derivative forms between Saurian and Bird than the isolated Pterodactyl, Archaeopteryx, Ichthyornis, and so forth, which we now have—that such a mathematician shall transform, by the variation of a few parameters, as in a dissolving view, one form into another, just as we transform one conic section into another.

Reverting now to Kirchhoff's words, we can come to some agreement regarding their import. Nothing can be built without building-stones, mortar, scaffolding, and a builder's skill. Yet assuredly the wish is well founded, that will show to posterity the complete structure in its finished form, bereft of unsightly scaffolding. It is the pure logical and æsthetic sense of the mathematician that speaks out of Kirchhoff's words. Modern expositions of physics aspire after his ideal; that, too, is intelligible. But it would be a poor didactic shift, for one whose business it was to train architects, to say: "Here is a splendid edifice; if thou wouldst really build, go thou and do likewise.

The barriers between the special sciences, which

make division of work and concentration possible, but which appear to us after all as cold and conventional restrictions, will gradually disappear. Bridge upon bridge is thrown over the gaps. Contents and methods, even of the remotest branches, are compared. When the Congress of Natural Scientists shall meet a hundred years hence, we may expect that they will represent a unity in a higher sense than is possible to-day, not in sentiment and aim alone, but in method also. In the meantime, this great change will be helped by our keeping constantly before our minds the fact of the intrinsic relationship of all research, which Kirchhoff characterised with such classical simplicity.

ON INSTRUCTION IN THE CLASSICS
AND THE SCIENCES.*

PERHAPS the most fantastic proposition that Maupertuis,† the renowned president of the Berlin
Academy, ever put forward for the approval of his

*An address delivered before the Congress of Delegates of the German
Realschulmännerverein, at Dortmund, April 16, 1886. The full title of the
address reads · "On the Relative Educational Value of the Classics and the
Mathematico-Physical Sciences in Colleges and High Schools."
   Although substantially contained in an address which I was to have made
at the meeting of Natural Scientists at Salzburg in 1881 (deferred on account
of the Paris Exposition), and in the Introduction to a course of lectures on
"Physical Instruction in Preparatory Schools," which I delivered in 1883, the
invitation of the German Realschulmännerverein afforded me the first opportunity of putting my views upon this subject before a large circle of readers.
Owing to the place and circumstances of delivery, my remarks apply of course,
primarily, only to German schools, but, with slight modifications, made in
this translation, are not without force for the institutions of other countries.
In giving here expression to a strong personal conviction formed long ago, it
is a matter of deep satisfaction to me to find that they agree in many points
with the views recently advanced in independent form by Paulsen (*Geschichte
des gelehrten Unterrichts*, Leipsic, 1885) and Frary (*La question du latin*,
Paris, Cerf, 1885). It is not my desire nor effort here to say much that is new,
but merely to contribute my mite towards bringing about the inevitable revolution now preparing in the world of elementary instruction. In the opinion
of experienced educationists the first result of that revolution will be to make
Greek and mathematics alternately optional subjects in the higher classes of
the German Gymnasium and in the corresponding institutions of other countries, as has been done in the splendid system of instruction in Denmark. The
gap between the German classical Gymnasium and the German Realgymnasium, or between classical and scientific schools generally, can thus be bridged
over, and the remaining inevitable transformations will then be accomplished
in relative peace and quiet. (Prague, May, 1886.)
   † Maupertuis, *Œuvres*, Dresden, 1752, p. 339.

contemporaries was that of founding a city in which, to instruct and discipline young students, only Latin should be spoken.   Maupertuis's Latin city remained an idle wish.   But for centuries Latin and Greek *institutions* exist in which our children spend a goodly portion of their days, and whose atmosphere constantly surrounds them, even when without their walls.

For centuries instruction in the ancient languages has been zealously cultivated.   For centuries its necessity has been alternately championed and contested. More strongly than ever are authoritative voices now raised against the preponderance of instruction in the classics and in favor of an education more suited to the needs of the time, especially for a more generous treatment of mathematics and the natural sciences.

In accepting your invitation to speak here on the relative educational value of the classical and the mathematico-physical sciences in colleges and high schools, I find my justification in the duty and the necessity laid upon every teacher of forming from his own experiences an opinion upon this important question, as partly also in the special circumstance that in my youth I was personally under the influence of school-life for only a short time, just previous to my entering the university, and had, therefore, ample opportunity to observe the effects of widely different methods upon my own person.

Passing, now, to a review of the arguments which the advocates of instruction in the classics advance,

and of what the adherents of instruction in the physical sciences in their turn adduce, we find ourselves in
rather a perplexing position with respect to the arguments of the first named. For these have been different at different times, and they are even now of a very
multifarious character, as must be where men advance,
in favor of an institution that exists and which they are
determined to retain at any cost, everything they can
possibly think of. We shall find here much that has
evidently been brought forward only to impress the
minds of the ignorant; much, too, that was advanced
in good faith and which is not wholly without foundation. We shall get a fair idea of the reasoning employed
by considering, first, the arguments that have grown
out of the historical circumstances connected with the
original introduction of the classics, and, lastly, those
which were subsequently adduced as accidental after-
thoughts.

*　　*　　*

Instruction in Latin, as Paulsen* has minutely
shown, was introduced by the Roman Church along
with Christianity. With the Latin language were also
transmitted the scant and meagre remnants of ancient
science. Whoever wished to acquire this ancient education, then the only one worthy of the name, for him
the Latin language was the only and indispensable
means; such a person had to learn Latin to rank
among educated people.

———————
*F. Paulsen, *Geschichte des gelehrten Unterrichts*, Leipsic, 1885.

The wide-spread influence of the Roman Church wrought many and various results. Among those for which all are glad, we may safely count the establishment of a sort of *uniformity* among the nations and of a regular international intercourse by means of the Latin language, which did much to unite the nations in the common work of civilisation, carried on from the fifteenth to the eighteenth century. The Latin language was thus long the language of scholars, and instruction in Latin the road to a liberal education—a shibboleth still employed, though long inappropriate.

For scholars as a class, it is to be regretted, perhaps, that Latin has ceased to be the medium of international communication. But the attributing of the loss of this function by the Latin language to its incapacity to accommodate itself to the numerous new ideas and conceptions which have arisen in the course of the development of science is, in my opinion, wholly erroneous. It would be difficult to find a modern scientist who had enriched science with as many new ideas as Newton has, yet Newton knew how to express those ideas very correctly and precisely in the Latin language. If this view were correct, it would also hold true of every living language. Originally every language has to adapt itself to new ideas.

It is far more likely that Latin was displaced as the literary vehicle of science by the influence of the nobility. By their desire to enjoy the fruits of literature and science, through a less irksome medium than

Latin, the nobility performed for the people at large an undeniable service. For the days were now past when acquaintance with the language and literature of science was restricted to a caste, and in this step, perhaps, was made the most important advance of modern times. To-day, when international intercourse is firmly established in spite of the many languages employed, no one would think of reintroducing Latin.*

The facility with which the ancient languages lend themselves to the expression of new ideas is evidenced by the fact that the great majority of our scientific ideas, as survivals of this period of Latin intercourse, bear Latin and Greek designations, while in great measure scientific ideas are even now invested with names from these sources. But to deduce from the existence and use of such terms the necessity of still learning Latin and Greek on the part of all who employ them is carrying the conclusion too far. All terms, appropriate and inappropriate,—and there are a large number of inappropriate and monstrous combinations in science,—rest on convention. The essential thing is, that people should associate with the sign the precise idea that is designated by it. It matters little whether a person can correctly derive the words *telegraph, tangent, ellipse, evolute,* etc., if the correct idea

---

* There is a peculiar irony of fate in the fact that while Leibnitz was casting about for a new vehicle of universal linguistic intercourse, the Latin language which still subserved this purpose the best of all, was dropping more and more out of use, and that Leibnitz himself contributed not the least to this result.

is present in his mind when he uses them. On the other hand, no matter how well he may know their etymology, his knowledge will be of little use to him if the correct idea is absent. Ask the average and fairly educated classical scholar to translate a few lines for you from Newton's *Principia*, or from Huygens's *Horologium*, and you will discover at once what an extremely subordinate rôle the mere knowledge of language plays in such things. Without its associated thought a word remains a mere sound. The fashion of employing Greek and Latin designations—for it can be termed nothing else—has a natural root in history; it is impossible for the practice to disappear suddenly, but it has fallen of late considerably into disuse. The terms *gas*, *ohm*, *Ampère*, *volt*, etc., are in international use, but they are not Latin nor Greek. Only the person who rates the unessential and accidental husk higher than its contents, can speak of the necessity of learning Latin or Greek for such reasons, to say nothing of spending eight or ten years on the task. Will not a dictionary supply in a few seconds all the information we wish on such subjects?*

*As a rule, the human brain is too much, and wrongly, burdened with things which might be more conveniently and accurately preserved in books where they could be found at a moment's notice. In a recent letter to me from Düsseldorf, Judge Hartwich writes:

"A host of words exist which are out and out Latin or Greek, yet are em-"ployed with perfect correctness by people of good education who never had "the good luck to be taught the ancient languages. For example, words like "'dynasty.'. . . The child learns such words as parts of the common stock of "speech, or even as parts of his mother-tongue, just as he does the words "'father,' 'mother,' 'bread,' 'milk.' Does the ordinary mortal know the ety-"mology of these Saxon words? Did it not require the almost incredible

It is indisputable that our modern civilisation took up the threads of the ancient civilisation, that at many points it begins where the latter left off, and that centuries ago the remains of the ancient culture were the only culture existing in Europe.   Then, of course, a classical education really was the liberal education, the higher education, the ideal education, for it was the *sole* education.   But when the same claim is now raised in behalf of a classical education, it must be uncompromisingly contested as bereft of all foundation.   For our civilisation has gradually attained its independence ; it has lifted itself far above the ancient civilisation, and has entered generally new directions of progress.   Its note, its characteristic feature, is the enlightenment that has come from the great mathematical and physical researches of the last centuries, and which has permeated not only the practical arts and industries but is also gradually finding its way into all fields of thought, including philosophy and history, sociology and linguistics.   Those traces of ancient views that are still discoverable in philosophy, law, art, and science, operate more as hindrances than helps, and will not long stand before the development of independent and more natural views.

"industry of the Grimms and other Teutonic philologists to throw the merest
"glimmerings of light upon the origin and growth of our own mother-tongue ?
"Besides, do not thousands of people of so-called classical education use
"every moment hosts of words of foreign origin whose derivation they do not
"know?   Very few of them think it worth while to look up such words in the
"dictionaries, although they love to maintain that people should study the
"ancient languages for the sake of etymology alone."

It ill becomes classical scholars, therefore, to regard themselves, at this day, as the educated class
par excellence, to condemn as uneducated all persons
who do not understand Latin and Greek, to complain
that with such people profitable conversations are not
to be carried on, etc.   The most delectable stories
have got into circulation, illustrative of the defective
education of scientists and engineers.   A renowned
inquirer, for example, is said to have once announced
his intention of holding a free course of university lectures, with the word "frustra"; an engineer who spent
his leisure hours in collecting insects is said to have
declared that he was studying "etymology."   It is
true, incidents of this character make us shudder or
smile, according to our mood or temperament.   But
we must admit, the next moment, that in giving way
to such feelings we have merely succumbed to a childish prejudice.   A lack of tact but certainly no lack of
education is displayed in the use of such half-under-
stood expressions.   Every candid person will confess
that there are many branches of knowledge about which
he had better be silent.   We shall not be so uncharitable as to turn the tables and discuss the impression
that classical scholars might make on a scientist or
engineer, in speaking of science.   Possibly many ludi-
crous stories might be told of them, and of far more
serious import, which should fully compensate for the
blunders of the other party.

The mutual severity of judgment which we have

here come upon, may also forcibly bring home to us
how really scarce a true liberal culture is. We may
detect in this mutual attitude, too, something of that
narrow, mediæval arrogance of caste, where a man
began, according to the special point of view of the
speaker, with the scholar, the soldier, or the nobleman.
Little sense or appreciation is to be found in it for the
common task of humanity, little feeling for the need of
mutual assistance in the great work of civilisation,
little breadth of mind, little truly liberal culture.

A knowledge of Latin, and partly, also, a knowledge of Greek, is still a necessity for the members of
a few professions by nature more or less directly concerned with the civilisations of antiquity, as for lawyers, theologians, philologists, historians, and generally for a small number of persons, among whom
from time to time I count myself, who are compelled
to seek for information in the Latin literature of the
centuries just past.* But that all young persons in
search of a higher education should pursue for this
reason Latin and Greek to such excess ; that persons
intending to become physicians and scientists should
come to the universities defectively educated, or even
miseducated ; and that they should be compelled to

* Standing remote from the legal profession I should not have ventured to
declare that the study of Greek was not necessary for the jurists; yet this
view was taken in the debate that followed this lecture by professional jurists
of high standing. According to this opinion, the preparatory education obtained in the German Realgymnasium would also be sufficient for the future
jurists and insufficient only for theologians and philologists. [In England and
America not only is Greek not necessary, but the law-Latin is so peculiar that
even persons of *good* classical education cannot understand it.—*Tr.*]

come only from schools that do *not* supply them with
the proper preparatory knowledge is going a little bit
too far.

*   *   *

After the conditions which had given to the study
of Latin and Greek their high import had ceased to
exist, the traditional curriculum, naturally, was retained.   Then, the different effects of this method of
education, good and bad, which no one had thought of
at its introduction, were realised and noted.   As natural, too, was it that those who had strong interests
in the preservation of these studies, from knowing no
others or from living by them, or for still other reasons, should emphasise the *good* results of such instruction.   They pointed to the good effects as if they
had been consciously aimed at by the method and could
be attained only through its agency.

One real benefit that students might derive from
a rightly conducted course in the classics would be
the opening up of the rich literary treasures of antiquity, and intimacy with the conceptions and views
of the world held by two advanced nations.   A person
who has read and understood the Greek and Roman
authors has felt and experienced more than one who is
restricted to the impressions of the present.   He sees
how men placed in different circumstances judge quite
differently of the same things from what we do to-day.
His own judgments will be rendered thus more independent.   Again, the Greek and Latin authors are indis-

putably a rich fountain of recreation, of enlightenment, and of intellectual pleasure after the day's toil, and the individual, not less than civilised humanity generally, will remain grateful to them for all time. Who does not recall with pleasure the wanderings of Ulysses, who does not listen joyfully to the simple narratives of Herodotus, who would ever repent of having made the acquaintance of Plato's Dialogues, or of having tasted Lucian's divine humor? Who would give up the glances he has obtained into the private life of antiquity from Cicero's letters, from Plautus or Terence? To whom are not the portraits of Suetonius undying reminiscences? Who, in fact, would throw away *any* knowledge he had once gained?

Yet people who draw from these sources only, who know only this culture, have surely no right to dogmatise about the value of some other culture. As objects of research for individuals, this literature is extremely valuable, but it is a different question whether it is equally valuable as the almost exclusive means of education of our youth.

Do not other nations and other literatures exist from which we ought to learn? Is not nature herself our first school-mistress? Are our highest models always to be the Greeks, with their narrow provinciality of mind, that divided the world into "Greeks and barbarians," with their superstitions, with their eternal questioning of oracles? Aristotle with his incapacity to learn from facts, with his word-science; Plato with his heavy, interminable dialogues, with his barren, at times childish, dialectics—are they unsurpassable ? * The Romans with their apathy, their pompous externality, set off by fulsome and bombastic phrases, with their narrow-minded, philistine philosophy, with their frenzied sensuality, with their cruel and bestial indulgence in animal and man baiting, with their outrageous maltreatment and plundering of their subjects—are they patterns worthy of imitation ? Or shall, perhaps, our science edify itself with the works of Pliny who cites midwives as authorities and himself stands on their point of view?

Besides, if an acquaintance with the ancient world really were attained, we might come to some settlement with the advocates of classical education. But it is words and forms, and forms and words only, that are supplied to our youth ; and even collateral subjects are forced into the strait-jacket of the same rigid method and made a science of words, sheer feats of mechanical memory. Really, we feel ourselves set back a thousand years into the dull cloister-cells of the Middle Ages.

This must be changed. It is possible to get ac-

* In emphasising here the weak sides of the writings of Plato and Aristotle, forced on my attention while reading them in German translations, I, of course, have no intention of underrating the great merits and the high historical importance of these two men. Their importance must not be measured by the fact that our speculative philosophy still moves to a great extent in their paths of thought. The more probable conclusion is that this branch has made very little progress in the last two thousand years. Natural science also was implicated for centuries in the meshes of the Aristotelian thought, and owes its rise mainly to having thrown off those fetters.

quainted with the views of the Greeks and Romans by a shorter road than the intellect-deadening process of eight or ten years of declining, conjugating, analysing, and extemporisation.   There are to-day plenty of educated persons who have acquired through good translations vivider, clearer, and more just views of classical antiquity than the graduates of our gymnasiums and colleges.*

For us moderns, the Greeks and the Romans are simply two objects of archæological and historical research like all others.   If we put them before our youth in fresh and living pictures, and not merely in words and syllables, the effect will be assured.   We derive a totally different enjoyment from the Greeks when we approach them after a study of the results of modern research in the history of civilisation.   We read many a chapter of Herodotus differently when we attack his works equipped with a knowledge of natural science, and with information about the stone age and the lake-dwellers.   What our classical institutions *pretend* to give can and actually will be given to our youth with much more fruitful results by competent *historical* instruction, which must supply, not names and numbers alone, nor the mere history of dynasties and wars, but be in every sense of the word a true history of civilisation.

* I would not for a moment contend that we derive exactly the same profit from reading a Greek author in a translation as from reading him in the original; but the difference, the excess of gain in the second case, appears to me, and probably will to most men who are not professional philologists, to be too dearly bought with the expenditure of eight years of valuable time.

The view still widely prevails that although all "higher, ideal culture," all extension of our view of the world, is acquired by philological and in a lesser degree by historical studies, still the mathematics and natural sciences should not be neglected on account of their usefulness. This is an opinion to which I must refuse my assent. It were strange if man could learn more, could draw more intellectual nourishment, from the shards of a few old broken jugs, from inscribed stones, or yellow parchments, than from all the rest of nature. True, man is man's first concern, but he is not his sole concern.

In ceasing to regard man as the centre of the world; in discovering that the earth is a top whirled about the sun, which speeds off with it into infinite space; in finding that in the fixed stars the same elements exist as on earth; in meeting everywhere the same processes of which the life of man is merely a vanishingly small part—in such things, too, is a widening of our view of the world, and edification, and poetry. There are here perhaps grander and more significant facts than the bellowing of the wounded Ares, or the charming island of Calypso, or the ocean-stream engirdling the earth. He only should speak of the relative value of these two domains of thought, of their poetry, who knows both.

The "utility" of physical science is, in a measure, only a *collateral* product of that flight of the intellect which produced science. No one, however, should underrate the utility of science who has shared in the realisation by modern industrial art of the Oriental world of fables, much less one upon whom those treasures have been poured, as it were, from the fourth dimension, without his aid or understanding.

Nor may we believe that science is useful only to the practical man. Its influence permeates all our affairs, our whole life; everywhere its ideas are decisive. How differently does the jurist, the legislator, or the political economist think, who knows, for example, that a square mile of the most fertile soil can support with the solar heat annually consumed only a definite number of human beings, which no art or science can increase. Many economical theories, which open new air-paths of progress, air-paths in the literal sense of the word, would be made impossible by such knowledge.

* * *

The eulogists of classical education love to emphasise the cultivation of taste which comes from employment with the ancient models. I candidly confess that there is something absolutely revolting in this to me. To form the taste, then, our youths must sacrifice ten years of their life ! Luxury takes precedence over necessity. Have the future generations, in the face of the difficult problems, the great social questions, which they must meet, and that with strengthened mind and heart, no more important duties to fulfil than these ?

But let us assume that this end were desirable. Can taste be formed by rules and precepts? Do not ideals of beauty change? Is it not a stupendous absurdity to force one's self artificially to admire things which, with all their historical interest, with all their beauty in individual points, are for the most part foreign to the rest of our thoughts and feelings, provided we have such of *our own*. A nation that is truly such, has its own taste and will not go to others for it. And every individual perfect man has his own taste.*

And what, after all, does this cultivation of taste consist in? In the acquisition of the personal literary style of a few select authors! What should we think of a people that would force its youth a thousand years from now, by years of practice, to master the tortuous or bombastic style of some successful lawyer or politician of to-day? Should we not justly accuse them of a woful lack of taste?

The evil effects of this imagined cultivation of the

---

* "The temptation," Judge Hartwich writes, "to regard the 'taste' of the "ancients as so lofty and unsurpassable appears to me to have its chief origin "in the fact that the ancients were unexcelled in the representation of the "nude. First, by their unremitting care of the human body they produced "splendid models; and secondly, in their gymnasiums and in their athletic "games they had these models constantly before their eyes. No wonder, then, "that their statues still excite our admiration! For the form, the ideal of the "human body has not changed in the course of the centuries. But with intel- "lectual matters it is totally different; they change from century to century, "nay, from decennium to decennium. It is very natural now, that people "should unconsciously apply what is thus so easily seen, namely, the works of "sculpture, as a universal criterion of the highly developed taste of the an- "cients—a fallacy against which people cannot, in my judgment, be too strongly "warned."

taste find expression often enough. The young *savant*
who regards the composition of a scientific essay as a
rhetorical exercise instead of a simple and unadorned
presentation of the facts and the truth, still sits unconsciously on the school-bench, and still unwittingly represents the point of view of the Romans, by whom the
elaboration of speeches was regarded as a serious scientific (!) employment.

*   *   *

Far be it from me to underrate the value of the development of the instinct of speech and of the increased
comprehension of our own language which comes from
philological studies. By the study of a foreign language, especially of one which differs widely from ours,
the signs and forms of words are first clearly distinguished from the thoughts which they express. Words
of the closest possible correspondence in different languages never coincide absolutely with the ideas they
stand for, but place in relief slightly different aspects
of the same thing, and by the study of language the
attention is directed to these shades of difference. But
it would be far from admissible to contend that the
study of Latin and Greek is the most fruitful and natural, let alone the *only*, means of attaining this end.
Any one who will give himself the pleasure of a few
hours' companionship with a Chinese grammar ; who
will seek to make clear to himself the mode of speech
and thought of a people who never advanced as far as
the analysis of articulate sounds, but stopped at the

analysis of syllables, to whom our alphabetical characters, therefore, are an inexplicable puzzle, and who
express all their rich and profound thoughts by means
of a few syllables with variable emphasis and position,
—such a person, perhaps, will acquire new, and extremely elucidative ideas upon the relation of language and thought.   But should our children, therefore, study Chinese?   Certainly not.   No more, then,
should they be burdened with Latin, at least in the
measure they are.

It is a beautiful achievement to reproduce a Latin
thought in a modern language with the maximum fidelity of meaning and expression — for the *translator*.
Moreover, we shall be very grateful to the translator
for his performance.   But to demand this feat of every
educated man, without consideration of the sacrifice of
time and labor which it entails, is unreasonable. And
for this very reason, as classical teachers admit, that
ideal is never perfectly attained, except in rare cases
with scholars possessed of special talents and great
industry.   Without slurring, therefore, the high importance of the study of the ancient languages as a
profession, we may yet feel sure that the instinct for
speech which is part of every liberal education can,
and must, be acquired in a different way.   Should we,
indeed, be forever lost if the Greeks had not lived before us?

The fact is, we must carry our demands further
than the representatives of classical philology.   We

must ask of every educated man a fair scientific conception of the nature and value of language, of the
formation of language, of the alteration of the meaning of roots, of the degeneration of fixed forms of
speech to grammatical forms, in brief, of all the main
results of modern comparative philology. We should
judge that this were attainable by a careful study of
our mother tongue and of the languages next allied to
it, and subsequently of the more ancient tongues from
which the former are derived. If any one object that
this is too difficult and entails too much labor, I should
advise such a person to place side by side an English,
a Dutch, a Danish, a Swedish, and a German Bible, and
to compare a few lines of them; he will be amazed at
the multitude of suggestions that offer themselves.*
In fact, I believe that a really progressive, fruitful, ra-
tional, and instructive study of languages can be conducted only on this plan. Many of my audience will
remember, perhaps, the bright and encouraging effect,
like that of a ray of sunlight on a gloomy day, which
the meagre and furtive remarks on comparative phi-

---

*English: "In the beginning God created the heaven and the earth.
"And the earth was without form and void; and darkness was upon the face
"of the deep. And the spirit of God moved upon the face of the waters."—
Dutch: "In het begin schiep God den hemel en de aarde. De aarde nu was
"woest en ledig, en duisternis was op den afgrond; en de Geest Gods zwefde
"op de wateren."—Danish: "I Begyndelsen skabte Gud Himmelen og Jor-
"den. Og Jorden var ode og tom, og der var morkt ovenover Afgrunden, og
"Guds Aand svoevede ovenover Vandene."—Swedish: "I begynnelsen ska-
"pade Gud Himmel och Jord. Och Jorden war öde och tom, och mörker war
"pa djupet, och Gods Ande swåfde öfwer wattnet."—German: "Am Anfang
"schuf Gott Himmel und Erde. Und die Erde war wüst und leer, und es war
"finster auf der Tiefe; und der Geist Gottes schwebte auf dem Wasser."

lology in Curtius's Greek grammar wrought in that
barren and lifeless desert of verbal quibbles.

*   *   *

The principal result obtained by the present method
of studying the ancient languages is that which comes
from the student's employment with their complicated
grammars.   It consists in the sharpening of the attention and in the exercise of the judgment by the practice of subsuming special cases under general rules,
and of distinguishing between different cases.   Ob-
viously, the same result can be reached by many other
methods ; for example, by difficult games of cards.
Every science, the mathematics and the physical sciences included, accomplish as much, if not more, in
this disciplining of the judgment.   In addition, the
matter treated by those sciences has a much higher intrinsic interest for young people, and so engages spontaneously their attention ; while on the other hand they
are elucidative and useful in other directions in which
grammar can accomplish nothing.

Who cares, so far as the matter of it is concerned,
whether we say *hominum* or *hominorum* in the genitive
plural, interesting as the fact may be for the philologist ?   And who would dispute that the intellectual
need of causal insight is awakened not by grammar
but by the natural sciences ?

It is not our intention, therefore, to gainsay in the
least the good influence which the study of Latin and
Greek grammar *also* exercises on the sharpening of the

judgment. In so far as the study of words as such must greatly promote lucidity and accuracy of expression, in so far as Latin and Greek are not yet wholly indispensable to many branches of knowledge, we willingly concede to them a place in our schools, but would demand that the disproportionate amount of time allotted to them, wrongly withdrawn from other useful studies, should be considerably curtailed. That in the end Latin and Greek will not be employed as the universal means of education, we are fully convinced. They will be relegated to the closet of the scholar or professional philologist, and gradually make way for the modern languages and the modern science of language.

Long ago Locke reduced to their proper limits the exaggerated notions which obtained of the close connexion of thought and speech, of logic and grammar, and recent investigators have established on still surer foundations his views. How little a complicated grammar is necessary for expressing delicate shades of thought is demonstrated by the Italians and French, who, although they have almost totally discarded the grammatical redundancies of the Romans, are yet not surpassed by the latter in accuracy of thought, and whose poetical, but especially whose scientific literature, as no one will dispute, can bear favorable comparison with the Roman.

Reviewing again the arguments advanced in favor of the study of the ancient languages, we are obliged

to say that in the main and as applied to the present,
they are wholly devoid of force. In so far as the
aims which this study theoretically pursues are still
worthy of attainment, they appear to us as altogether
too narrow, and are surpassed in this only by the
means employed. As almost the sole, indisputable result of this study we must count the increase of the
student's skill and precision in expression. One inclined to be uncharitable might say that our gymnasiums and classical academies turn out men who can
speak and write, but, unfortunately, have little to write
or speak about. Of that broad, liberal view, of that
famed universal culture, which the classical curriculum
is supposed to yield, serious words need not be lost.
This culture might, perhaps, more properly be termed
the contracted or lopsided culture.

*   *   *

While considering the study of languages we threw
a few side glances at mathematics and the natural sciences.   Let us now inquire whether these, as branches
of study, cannot accomplish much that is to be attained
in no other way.   I shall meet with no contradiction
when I say that without at least an elementary mathematical and scientific education a man remains a total
stranger in the world in which he lives, a stranger in
the civilisation of the time that bears him.   Whatever
he meets in nature, or in the industrial world, either
does not appeal to him at all, from his having neither

eye nor ear for it, or it speaks to him in a totally unintelligible language.

A real understanding of the world and its civilisation, however, is not the only result of the study of mathematics and the physical sciences. Much more essential for the preparatory school is the *formal* cultivation which comes from these studies, the strengthening of the reason and the judgment, the exercise of the imagination. Mathematics, physics, chemistry, and the so-called descriptive sciences are so much alike in this respect, that, apart from a few points, we need not separate them in our discussion.

Logical sequence and continuity of ideas, so necessary for fruitful thought, are *par excellence* the results of mathematics; the ability to follow facts with thoughts, that is, to observe or collect experiences, is chiefly developed by the natural sciences. Whether we notice that the sides and the angles of a triangle are connected in a definite way, that an equilateral triangle possesses certain definite properties of symmetry, or whether we notice the deflexion of a magnetic needle by an electric current, the dissolution of zinc in diluted sulphuric acid, whether we remark that the wings of a butterfly are slightly colored on the under, and the fore-wings of the moth on the upper, surface : indiscriminately here we proceed from *observations*, from individual acts of immediate intuitive knowledge. The field of observation is more restricted and lies closer at hand in mathematics; it is more varied and broader but more difficult to compass in the natural sciences. The essential thing, however, is for the student to learn to make observations in all these fields. The philosophical question whether our acts of knowledge in mathematics are of a special kind is here of no importance for us. It is true, of course, that the observation can be practised by languages also. But no one, surely, will deny, that the concrete, living pictures presented in the fields just mentioned possess different and more powerful attractions for the mind of the youth than the abstract and hazy figures which language offers, and on which the attention is certainly not so spontaneously bestowed, nor with such good results.*

Observation having revealed the different properties of a given geometrical or physical object, it is discovered that in many cases these properties *depend* in some way upon one another. This interdependence of properties (say that of equal sides and equal angles at the base of a triangle, the relation of pressure to motion,) is nowhere so distinctly marked, nowhere is the necessity and permanency of the interdependence so plainly noticeable, as in the fields mentioned. Hence the continuity and logical consequence of the ideas which we acquire in those fields. The relative simplicity and perspicuity of geometrical and physical relations supply here the conditions of natural and

---

* Compare Herzen's excellent remarks, *De l'enseignement secondaire dans la Suisse romande.* Lausanne, 1886.

easy progress. Relations of equal simplicity are not met with in the fields which the study of language opens up. Many of you, doubtless, have often wondered at the little respect for the notions of cause and effect and their connexion that is sometimes found among professed representatives of the classical studies. The explanation is probably to be sought in the fact that the analogous relation of motive and action familiar to them from their studies, presents nothing like the clear simplicity and determinateness that the relation of cause and effect does.

That perfect mental grasp of all possible cases, that economical order and organic union of the thoughts which comes from it, which has grown for every one who has ever tasted it a permanent need which he seeks to satisfy in every new province, can be developed only by employment with the relative simplicity of mathematical and scientific investigations.

When a set of facts comes into apparent conflict with another set of facts, and a problem is presented, its solution consists ordinarily in a more refined distinction or in a more extended view of the facts, as may be aptly illustrated by Newton's solution of the problem of dispersion. When a new mathematical or scientific fact is *demonstrated*, or *explained*, such demonstration also rests simply upon showing the connexion of the new fact with the facts already known ; for example, that the radius of a circle can be laid off as chord exactly six times in the circle is explained or proved by dividing the regular hexagon inscribed in the circle into equilateral triangles. That the quantity of heat developed in a second in a wire conveying an electric current is quadrupled on the doubling of the strength of the current, we explain from the doubling of the fall of the potential due to the doubling of the current's intensity, as also from the doubling of the quantity flowing through, in a word, from the quadrupling of the work done. In point of principle, explanation and direct proof do not differ much.

He who solves scientifically a geometrical, physical, or technical problem, easily remarks that his procedure is a *methodical* mental quest, rendered possible by the economical order of the province—a simplified purposeful quest as contrasted with unmethodical, unscientific guess-work. The geometer, for example, who has to construct a circle touching two given straight lines, casts his eye over the relations of symmetry of the desired construction, and seeks the centre of his circle solely in the line of symmetry of the two straight lines. The person who wants a triangle of which two angles and the sum of the sides are given, grasps in his mind the determinateness of the form of this triangle and restricts his search for it to a certain group of triangles of the *same form*. Under very different circumstances, therefore, the simplicity, the intellectual perviousness, of the subject-matter of mathematics and natural science is felt, and promotes both the discipline and the self-confidence of the reason.

Unquestionably, much more will be attained by instruction in the mathematics and the natural sciences than now is, when more natural methods are adopted. One point of importance here is that young students should not be spoiled by premature abstraction, but should be made acquainted with their material from living pictures of it before they are made to work with it by purely ratiocinative methods. A good stock of geometrical experience could be obtained, for example, from geometrical drawing and from the practical construction of models. In the place of the unfruitful method of Euclid, which is only fit for special, restricted uses, a broader and more conscious method must be adopted, as Hankel has pointed out.* Then, if, on reviewing geometry, and after it presents no substantial difficulties, the more general points of view, the principles of scientific method are placed in relief and brought to consciousness, as Von Nagel,† J. K. Becker,‡ Mann,§ and others have well done, fruitful results will be surely attained. In the same way, the subject-matter of the natural sciences should be made familiar by pictures and experiment before a profounder and reasoned grasp of these subjects is attempted. Here the emphasis of the more general points of view is to be postponed.

Before my present audience it would be superfluous

* *Geschichte der Mathematik*, Leipsic, 1874.
† *Geometrische Analyse*, Ulm, 1886.
‡ In his text-books of elementary mathematics.
§ *Abhandlungen aus dem Gebiete der Mathematik*, Würzburg, 1883.

for me to contend further that mathematics and natural science are justified constituents of a sound education,—a claim that even philologists, after some
resistance, have conceded. Here I may count upon
assent when I say that mathematics and the natural
sciences pursued alone as means of instruction yield a
richer education in matter and form, a more general
education, an education better adapted to the needs
and spirit of the time,—than the philological branches
pursued alone would yield.

But how shall this idea be realised in the curricula
of our intermediate educational institutions ? It is unquestionable in my mind that the German *Realschulen*
and *Realgymnasien*, where the exclusive classical course
is for the most part replaced by mathematics, science,
and modern languages, give the *average* man a more
timely education than the gymnasium proper, although
they are not yet regarded as fit preparatory schools for
future theologians and professional philologists.   The
German gymnasiums are too one-sided.   With these
the first changes are to be made ; of these alone we
shall speak here.   Possibly a *single* preparatory school,
suitably planned, might serve all purposes.

Shall we, then, in our gymnasiums fill out the hours
of study which stand at our disposal, or are still to be
wrested from the classicists, with as great and as varied a quantity of mathematical and scientific matter
as possible ? Expect no such proposition from me.
No one will suggest such a course who has himself

been actively engaged in scientific thought. Thoughts
can be awakened and fructified as a field is fructified
by sunshine and rain. But thoughts cannot be jug-
gled out and worried out by heaping up materials and
the hours of instruction, nor by any sort of precepts :
they must grow naturally of their own free accord.
Furthermore, thoughts cannot be accumulated beyond
a certain limit in a single head, any more than the produce of a field can be increased beyond all limits.

I believe that the amount of matter necessary for a
useful education, such as should be offered to *all* the
pupils of a preparatory school, is very small. If I had
the requisite influence, I should, in all composure, and
fully convinced that I was doing what was best, first
greatly curtail in the lower classes the amount of matter in both the classical and the scientific courses ; I
should cut down considerably the number of the school
hours and the work done outside the school. I am
not with many teachers of opinion that ten hours work
a day for a child is not too much. I am convinced
that the mature men who offer this advice so lightly
are themselves unable to give their attention success-
fully for as long a time to any subject that is new to
them, (for example, to elementary mathematics or
physics,) and I would ask every one who thinks the
contrary to make the experiment upon himself. Learning and teaching are not routine office-work that can
be kept up mechanically for long periods. But even
such work tires in the end. If our young men are

not to enter the universities with blunted and impov-
erished minds, if they are not to leave in the preparatory schools their vital energy, which they should
there gather, great changes must be made.   Waiving
the injurious effects of overwork upon the body, the
consequences of it for the mind seem to me positively
dreadful.

I know of nothing more terrible than the poor creatures who have learned too much.   Instead of that
sound powerful judgment which would probably have
grown up if they had learned nothing, their thoughts
creep timidly and hypnotically after words, principles,
and formulæ, constantly by the same paths.   What
they have acquired is a spider's web of thoughts too
weak to furnish sure supports, but complicated enough
to produce confusion.

But how shall better methods of mathematical and
scientific education be combined with the decrease of
the subject-matter of instruction ?   I think, by aban-
doning systematic instruction altogether, at least in so
far as that is required of *all* young pupils.   I see no
necessity whatever that the graduates of our high
schools and preparatory schools should be little philologists, and at the same time little mathematicians,
physicists, and botanists ; in fact, I do not see the possibility of such a result.   I see in the endeavor to attain this result, in which every instructor seeks for his
own branch a place apart from the others, the main
mistake of our whole system.   I should be satisfied if

every young student could come into living contact
with and pursue to their ultimate logical consequences
merely a *few* mathematical or scientific discoveries.
Such instruction would be mainly and naturally associated with selections from the great scientific classics.
A few powerful and lucid ideas could thus be made
to take root in the mind and receive thorough elaboration. This accomplished, our youth would make a
different showing from what they do to-day.*

What need is there, for example, of burdening the
head of a young student with all the details of botany?
The student who has botanised under the guidance of
a teacher finds on all hands, not indifferent things, but
known or unknown things, by which he is stimulated,
and his gain made permanent. I express here, not my
own, but the opinion of a friend, a practical teacher.
Again, it is not at all necessary that all the matter that
is offered in the schools should be learned. The best
that we have learned, that which has remained with
us for life, outlived the test of examination. How can
the mind thrive when matter is heaped on matter, and
new materials piled constantly on old, undigested materials? The question here is not so much that of the
accumulation of positive knowledge as of intellectual

* My idea here is an appropriate selection of readings from Galileo, Huygens, Newton, etc. The choice is so easily made that there can be no question of difficulties. The contents would be discussed with the students, and
the original experiments performed with them. Those scholars alone should
receive this instruction in the upper classes who did not look forward to sys-
tematical instruction in the physical sciences. I do not make this proposition
of reform here for the first time. I have no doubt, moreover, that such radical
changes will only be slowly introduced.

discipline. It seems also unnecessary that *all* branches should be treated at school, and that exactly the same studies should be pursued in all schools. A single philological, a single historical, a single mathematical, a single scientific branch, pursued as common subjects of instruction for all pupils, are sufficient to accomplish all that is necessary for the intellectual development. On the other hand, a wholesome mutual stimulus would be produced by this greater variety in the positive culture of men. Uniforms are excellent for soldiers, but they will not fit heads. Charles V. learned this, and it should never be forgotten. On the contrary, teachers and pupils both need considerable latitude, if they are to yield good results.

With John Karl Becker I am of the opinion that the utility and amount for individuals of every study should be precisely determined. All that exceeds this amount should be unconditionally banished from the lower classes. With respect to mathematics, Becker,* in my judgment, has admirably solved this question.

With respect to the upper classes the demand assumes a different form. Here also the amount of matter obligatory on all pupils ought not to exceed a certain limit. But in the great mass of knowledge that a young man must acquire to-day for his profession it is no longer just that ten years of his youth should be wasted with mere preludes. The upper classes should supply a truly useful preparation for the professions,

* *Die Mathematik als Lehrgegenstand des Gymnasiums*, Berlin, 1883.

and should not be modelled upon the wants merely of future lawyers, ministers, and philologists. Again, it would be both foolish and impossible to attempt to prepare the same person properly for all the different professions. In such case the function of the schools would be, as Lichtenberg feared, simply to select the persons best fitted for being drilled, whilst precisely the finest special talents, which do not submit to indiscriminate discipline, would be excluded from the contest. Hence, a certain amount of liberty in the choice of studies must be introduced in the upper classes, by means of which it will be free for every one who is clear about the choice of his profession to devote his chief attention either to the study of the philologico-historical or to that of the mathematico-scientific branches. Then the matter now treated could be retained, and in some branches, perhaps, judiciously extended,* without burdening the scholar with many branches or increasing the number of the hours of study. With more homogeneous work the student's capacity for work increases, one part of his labor supporting the other instead of obstructing it. If, however, a young man should subsequently choose a different profession, then it is *his* business to make up what he has lost. No

---

* Wrong as it is to burden future physicians and scientists with Greek for the sake of the theologians and philologists, it would be just as wrong to compel theologians and philologists, on account of the physicians, to study such subjects as analytical geometry. Moreover, I cannot believe that ignorance of analytical geometry would be a serious hindrance to a physician that was otherwise well versed in quantitative thought. No special advantage generally is observable in the graduates of the Austrian gymnasiums, all of whom have studied analytical geometry. [Refers to an assertion of Dubois-Reymond.]

harm certainly will come to society from this change, nor could it be regarded as a misfortune if philologists and lawyers with mathematical educations or physical scientists with classical educations should now and then appear.

*   *   *

The view is now wide-spread that a Latin and Greek education no longer meets the general wants of the times, that a more opportune, a more "liberal" education exists.   The phrase, "a liberal education," has been greatly misused.   A truly liberal education is unquestionably very rare.   The *schools* can hardly offer such ; at best they can only bring home to the student the necessity of it.   It is, then, his business to acquire, as best he can, a more or less liberal education.   It would be very difficult, too, at any one time to give a definition of a "liberal" education which would satisfy every one, still more difficult to give one which would hold good for a hundred years.   The educational ideal, in fact, varies much.   To one, a knowledge of classical antiquity appears not too dearly bought "with early death."   We have no objection to this person, or to those who think like him, pursuing their ideal after their own fashion.   But we may certainly protest strongly against the realisation of such ideals on our own children.   Another,—Plato, for example,—puts men ignorant of geometry on a level with animals.*

---

*Compare M. Cantor, *Geschichte der Mathematik*, Leipsic, 1880, Vol. I, p. 193.

If such narrow views had the magical powers of the sorceress Circe, many a man who perhaps justly thought himself well educated would become conscious of a not very flattering transformation of himself. Let us seek, therefore, in our educational system to meet the wants of the present, and not establish prejudices for the future.

But how does it come, we must ask, that institutions so antiquated as the German gymnasiums could subsist so long in opposition to public opinion? The answer is simple. The schools were first organised by the Church; since the Reformation they have been in the hands of the State. On so large a scale, the plan presents many advantages. Means can be placed at the disposal of education such as no private source, at least in Europe, could furnish. Work can be conducted upon the same plan in many schools, and so experiments made of extensive scope which would be otherwise impossible. A single man with influence and ideas can under such circumstances do great things for the promotion of education.

But the matter has also its reverse aspect. The party in power works for its own interests, uses the schools for its special purposes. Educational competition is excluded, for all successful attempts at improvement are impossible unless undertaken or permitted by the State. By the uniformity of the people's education, a prejudice once in vogue is permanently established. The highest intelligences, the strongest

wills cannot overthrow it suddenly.   In fact, as everything is adapted to the view in question, a sudden
change would be physically impossible.   The two
classes which virtually hold the reins of power in the
State, the jurists and theologians, know only the one-
sided, predominantly classical culture which they have
acquired in the State schools, and would have this culture alone valued.   Others accept this opinion from
credulity; others, underestimating their true worth for
society, bow before the power of the prevalent opinion ; others, again, affect the opinion of the ruling
classes even against their better judgment, so as to
abide on the same plane of respect with the latter.   I
will make no charges, but I must confess that the de-
portment of medical men with respect to the question
of the qualification of graduates of your *Realschulen*
has frequently made that impression upon me.   Let
us remember, finally, that an influential statesman,
even within the boundaries which the law and public
opinion set him, can do serious harm to the cause
of education by considering his own one-sided views
infallible, and in enforcing them recklessly and incon-
siderately—which not only *can* happen, but has, re-
peatedly, happened.*   The monopoly of education by
the State† thus assumes in our eyes a somewhat different aspect.   And to revert to the question above asked,
there is not the slightest doubt that the German gym-

*Compare Paulsen, *l. c.*, pp. 607, 688.
† It is to be hoped that the Americans will jealously guard their schools
and universities against the influence of the State.

nasiums in their present form would have ceased to exist long ago if the State had not supported them.

All this must be changed. But the change will not be made of itself, nor without our energetic interference, and it will be made slowly. But the path is marked out for us, the will of the people must acquire and exert upon our school legislation a greater and more powerful influence. Furthermore, the questions at issue must be publicly and candidly discussed that the views of the people may be clarified. All who feel the insufficiency of the existing *régime* must combine into a powerful organisation that their views may acquire impressiveness and the opinions of the individual not die away unheard.

I recently read, gentlemen, in an excellent book of travels, that the Chinese speak with unwillingness of politics. Conversations of this sort are usually cut short with the remark that they may bother about such things whose business it is and who are paid for it. Now it seems to me that it is not only the business of the State, but a very serious concern of all of us, how our children shall be educated in the public schools at *our* cost.

INDEX.

INDEX.

307

et seq.; notions, energy and entropy are, 178; units, the building-stones of the physicist, 253.

Metronomes, experiment with, 41.

Meyer, Lothar, his periodical series, 256.

Middle Ages, 243, 270.

Mill, John Stuart, 230.

Mill-wheel, doing work, 161.

Mimetic reproduction of facts in thought, 193.

Mimicking facts in thought, 189.

Minor and major keys in music, 100 et seq.

Mirror, symmetrical reversion of objects in, 92 et seq.

Miserly mercantile principle at the basis of science, 15.

Moat, child looking into, 208.

Modern scientists, adherents of the mechanical philosophy, 188.

Molecular theories, 104.

Molecules, 203, 207.

Molière, 234.

Momentum, 184.

Monocular vision, 98.

Monotheism of the Christians and Jews, 187.

Montagues and Capulets, 87.

Moon, eclipse of, 219; lightness of bodies on, 4; the study of the, 90.

Mosaic of thought, 192.

Motion, a perpetual, 181; quantity of, 184; the Eleatics on, 158; Wundt on, 158; the Herbartians on, 158.

Motions, natural and violent, 226; their familiar character, 157.

Mountains of the earth, would crumble if very large, 3; weight of bodies on, 112.

Mozart, 44.

Multiplication-table, 195.

Multiplier, 132.

Music, band of, its *tempo* accelerated and retarded, 53; the principle of repetition in, 99 et seq.; its notation, mathematically illustrated, 103–104.

Musical notes, reversion of, 101 et seq.; their economy, 192.

Musical scale, a species of one-dimensional space, 105.

Mystery, in physics, 222; science can dispense with, 189.

Mysticism, numerical, 33; in the principle of energy, 184.

Mythology, the mechanical, of philosophy, 207.

Nagel, von, 285.

Napoleon, picture representing the tomb of, 36.

Natural constants, 193.

Natural law, a, not contained in the conformity of the energies, 175.

Natural laws, abridged descriptions, 193; likened to type, 193.

Natural motions, 225.

Natural selection in scientific theories, 63, 218.

Nature, experience the well-spring of all knowledge of, 181; fashions of, 64; first knowledge of, instinctive, 189; general interconnexion of, 182; has many sides, 217; her forces compared to purposes, 14–15; likened to a good man of business, 15; the economy of her actions, 15; how it appears to other animals, 83 et seq.; inquiry of, viewed as a torture, 48–49; view of, as something designedly concealed from man, 49; like a covetous tailor, 9–10; magic powers of, 189; our view of, modified by binocular vision, 82; the experimental method a questioning of, 48.

Negro hamlet, the science of a, 237.

Neptune, prediction and discovery of the planet, 29.

Newton, describes polarisation, 242; expresses his wealth of thought in Latin, 262; his discovery of gravitation, 225 et seq.; his solution of dispersion, 283; his principle of the equality of pressure and counterpressure, 191; his view of light, 227–228; on absolute time, 204; selections from his works for use in instruction, 289.

Nobility, displace Latin, 263.

Notation, musical, mathematically illustrated, 103–104.

Novel of a cockchafer, 86.
Numbers, economy of, 195; the simple natural, their connexion with consonance, 32.
Numerical mysticism, 33.
Nursery, the questions of the, 199.

Observation, in science, 261,
Ocean-stream, 272.
Oettingen, von, 103.
Ohm, on electric currents, 249.
Ohm, the word, 264.
Oil, alcohol, water, and, employed in Plateau's experiments, 4; free mass of, assumes the shape of a sphere, 12; geometrical figures of, 5 et seq.
One-eyed people, vision of, 98.
Ophthalmoscope, 18.
Optic nerves, 96.
Optimism and pessimism, 234.
Order of physics, 197.
Organ, bellows of an, 135.
Organic nature, results of Darwin's studies of, 215 et seq. See Adaptation and Heredity.
Oriental world of fables, 273.
Oscillation, centre of, 147 et seq.
Ostwald, 172.
Overtones, 38, 40.

Painted things, the difference between real and, 68.
Palestrina, 44.
Parameter, 257.
Particles, smallest, 104.
Paulsen, 259, 261, 294.
Pearls of life, strung on the individual as on a thread, 234–235.
Pencil surpasses the mathematician in intelligence, 196.
Pendulum, motion of a, 144 et seq.; increased motion of, due to slight impulses, 21; electrical, 110.
Periodical, changes, 181; series, 256.
Permanent, changes, 181, 199; elements of the world, 194.
Perpetual motion, a, 181; defined, 139; impossibility of, 139 et seq.; the principle of the excluded, 140 et seq.; excluded from general physics, 162.

Personality, its nature, 234–235.
Perspective, 76 et seq.; contraction of, 74 et seq.; distortion of, 77.
Pessimism and optimism, 234.
Pharaohs, 85.
Phenomenology, a universal physical, 250.
Philistine, modes of thought of the, 223.
Philology, comparison in, 239.
Philosopher, an ancient, on the moral and physical sciences, 89.
Philosophy, its character at all times, 186; mechanical, 155 et seq., 188, 207.
Phonetic alphabets, their economy, 192.
Photography, stupendous advances of, 74.
Physical, concepts, fetishism in our, 187; ideas and principles, their nature, 204; inquiry, the economical nature of, 186; research, object of, 207, 209.
Physical phenomena, as mechanical phenomena, 182; relations between, 205.
Physico-mechanical view of the world, 187, 188, 207, 155 et seq.
Physics, compared to a well-kept household, 197; economical experience, 197; the principles of, descriptive, 199; the methods of, 209: its method characterised, 211; comparison in, 239; the facts of, qualitatively homogeneous, 255; how it began, 37; helped by psychology, 104; study of its own character, 189; the goal of, 207, 209.
Physiological psychology, its methods, 211 et seq.
Physiology, its scope, 212.
Piano, its mirrored counterpart, 100 et seq.; used to illustrate the facts of sympathetic vibration, 25 et seq.
Piano-player, a speaker compared to, 192.
Picture, physical, a, 110.
Pillars of Corti, 19.
Places, heavy bodies seek their, 224 et seq.

Printed in the United States
By Bookmasters